Development and Diffusion of the Nuclear Power Reactor: A Comparative Analysis

• A Rand Graduate Institute Book •

Development and Diffusion of the Nuclear Power Reactor: A Comparative Analysis

PETER DELEON

Ballinger Publishing Company • Cambridge, Massachusetts
A Subsidiary of Harper & Row, Publishers, Inc.

This book is printed on recycled paper.

Library of Congress Cataloging in Publication Data

DeLeon, Peter.

A comparative analysis of high technology programs.

Originally presented as the author's thesis, Rand Graduate Institute for Policy Studies, 1978.
Bibliography: p.
Includes index.
1. Nuclear engineering—Research. 2. Nuclear reactors. 3. Technology and state. I. Title.
TK9153.D44 1979 338.4'7'62148 79-12988
ISBN 0-88410-682-9

International Standard Book Number: 0-88410-682-9

Library of Congress Catalog Card Number: 79-12988

Printed in the United States of America

Contents

List of Tables

tables

List of Figures

figures

Glossary

AEA	=	Atomic Energy Authority
AEC	=	Atomic Energy Commission
AECL	=	Atomic Energy of Canada Limited
AEG	=	Allegemeine Elektrizitätsgellschaft
AGR	=	Advanced Gas Reactor
AKB	=	Atomkraftwerk Bayern
APG	=	Atomic Power Group
AVR	=	Arbeitgemeinschaft Versuchreaktor
BNDC	=	British Nuclear Design and Construction
BWR	=	Boiling Water Reactor
CANDU	=	Canadian Deuterium Uranium (reactor)
CEA	=	Commissariat à l'Energie
CEGB	=	Central Electricity Generating Board
D_2O	=	Deuterium Oxide ("heavy water")
DAK	=	Deutsche Atomkommission
EBR	=	Experimental Breeder Reactor
EdF	=	Electricité de France
ENEA	=	European Nuclear Energy Agency
FBR	=	Fast Breeder Reactor
GE	=	General Electric

Glavatom = Chief Directorate to the Council of Ministers for the Utilization of Atomic Energy
HTGR = High Temperature Gas (cooled) Reactor
HWR = Heavy Water (cooled and moderated) Reactor
IRC = Industrial Reorganization Corporation
JCAE = Joint Committee on Atomic Energy
KwHr = Kilowatt-hours
LWGR = Light Water (cooled) Graphite (moderated) Reactor
LWR = Light Water Reactor
MAE = Ministry for Atomic Energy (adopted for consistency when referring to West German ministry for atomic energy)

MW(e) = Megawatts electricity
MW(th) = Megawatts thermal
KWU = Kernkraftwerk Union
MZFR = Mehrzweckforschungsreaktor (a German PHWR)
NPD = Nuclear Power Demonstration (Canadian reactor)
NPT = Non-Proliferation Treaty
ORGEL = Organique Eau Lourde (reactor)
PHWR = Pressurized Heavy Water Reactor (also PHW)
PEON = Commission Consultative pour la Production d'Electricité Nucléaire
PRDP = Power Reactor Demonstration Program
PRT = Personal Rapid Transit
PWR = Pressurized Water Reactor

R&D = Research and Development
RD&D = Research, Development, and Demonstration
RWE = Rheinisch-Westphälisches Elektrizitätswerk
SGHWR = Steam-Generating Heavy Water Reactor
SST = Supersonic Transport
TDS = Technology Delivery System
TNPG = The Nuclear Planning Group
UPC = United Power Group

Preface

This book was originally submitted as a thesis in partial requirement for the degree of Doctor of Philosophy in Policy Analysis offered by The Rand Graduate Institute. It represents an outgrowth of a Rand Corporation study sponsored by the National Science Foundation on the history of the light water reactor program.[1]

The comparative analysis of six national nuclear reactor programs should be of interest to a large community of scholars and practitioners, including those dealing with energy policy, technology assessment, government research and development of civilian technologies, comparative policy analysis, and the interaction between organizations and emerging technologies. Although the evidence is not sufficient to support or validate a full-blown theory of technology development, some attempt is made to suggest possible avenues for future theoretical exploration in that vein.

[1] The study produced four reports, the principal being Robert L. Perry et al., *Development and Commercialization of the Light Water Reactor, 1946-1976*, The Rand Corporation, R-2180-NSF, June 1977.

Acknowledgments

By its very nature, a dissertation has many contributors. I am happy to acknowledge them with the clear understanding that I can never quite repay them.

Robert L. Perry, Rand's director of the National Science Foundation history of the light water reactor, originally asked me to investigate foreign reactor programs; his early and continued support as the project grew into a dissertation was invaluable. Rand colleagues Nancy Nimitz and Sidney Siegel read and commented on an earlier draft of the narrative chapters (Chapters 3 to 7).

My dissertation committee at The Rand Graduate Institute was a constant source of ideas, clarifications, and encouragement. Upon reflection, I am not sure which of these benefits was the most important. Albert P. Williams assumed the dual and demanding roles of thesis chairman and personal friend. His suggestions were invariably sound; the confidences he expressed during the "dog days of the dissertation" were especially appreciated. James L. Foster consistently pointed me toward a clearer presentation of the analysis. Paul Y. Hammond of the University of Pittsburgh encouraged the study's comparative thrust. Walter S. Baer was instrumental at an early stage of the dissertation.

The Rand Graduate Institute provided support for the preparation of the manuscript. My secretary, Mildred Vukmanic, performed heroic typing feats in translating my drafts into legible copy; I suspect she is more relieved to see the dissertation completed than any other person. Helen Turin edited the final text with her accustomed skill.

Having indicted all the above people, I must now absolve them of their responsibilities and assume the full burden of guilt and opprobrium for any and all errors of fact or interpretation that might remain.

Part I:
Advanced Technology and Public Policy

1

Introduction

The growth, effects, and costs of high risk, advanced technologies have become increasingly pressing topics for public attention over the past decade. The previously unquestioned benefits assumed to accrue from advanced technology and, concomitantly, its more recently perceived economic and social costs[1] are now closely examined, particularly in areas in which the government must invest heavily to develop the technology. Examples are easy to cite: various forms of nuclear and thermonuclear energy technologies; mass transit systems; supersonic airliners; and on a lesser scale, some types of medical and biological research and technology.

The development and application of advanced technologies

[1] Although H.G. Wells warned about the potential dangers of "rampant" technology at the beginning of the twentieth century, it was not until the 1960s that the perceived negative effect of modern technology on the individual received widespread and respected currency. A widely read example is Alvin Toffler, *Future Shock* (New York: Random House, 1970).

with government sponsorship have generated three general types of public policy questions. The first and most general addresses the long-assumed efficacy of technology as it applies to social problems and asks if the notion of a technological "fix" is valid for important social policy issues.[2] For example, "If America can put a man on the moon, why can't it get rid of urban blight?"[3] Policymakers no longer equate technological progress with societal betterment.[4] Nor are they unquestioningly accepting technological solutions to social problems; they recognize the contextual issues as much too complex and even counterintuitive to be susceptible to the simple application of new technologies. Policymakers are demanding more complete contextual analyses of new technologies and their effects; these emphasize the attendant economic, environmental, and social costs, with a greater weight placed on noneconomic costs than in earlier instances. The governor of New York has publicly opposed granting landing rights to the advanced supersonic transport, the Concorde, at New York's Kennedy International Airport because, he argued, its landing would represent a measurable and significant decrement in the quality of life for those living within the Concorde's landing pattern and return very little of commensurate value. His position was particularly striking because it stood in clear opposition to the policy preferences stated by the relevant federal officials, including the president himself.

A second set of policy questions addresses the proper role of government in the development and dissemination of advanced technologies. Indeed, should the government even become involved? This set of questions has often been argued from a perspective of economics. Some economists frame their analyses in terms of government interventions to correct market failures,[5]

---

[2] An example of technology applied to social problems is provided by Amitai Etzioni and Richard Remp, "Technological 'Shortcuts' to Social Changes," *Science* 175, 4017 (January 6, 1972):31-38.

[3] A thoughtful examination of this query is Richard R. Nelson, "Intellectualizing About the Moon-Ghetto Metaphor: A Study of the Current Malaise of Rational Analysis of Social Problems," *Policy Sciences* 5, 4 (December 1974):375-414.

[4] One of the major rationales behind the "technology assessment" movement is to appreciate this relationship; see Joseph P. Martino, *Technological Forecasting for Decisionmaking* (New York: American Elsevier, 1972), for a description of the various methodologies.

[5] See George Eads, "U.S. Government Support for Civilian Technology:

the economic return of advanced technology to the population, or whether technology research and development should be treated as a "public good."[6] Others have assumed that the government has an appropriate role to play but question when and how the government should turn the fruit of its research and development over to private industry; in other words, at what point should a market mechanism be introduced that would prevent the government from "forcing" an economically unviable technology upon the society?[7] Still another school of thought assumes that some sort of positive relationship among advanced technology, economic growth, and general social welfare can be established, and it then asks how the government investment in specific, selected technologies should be structured.[8]

A third general question is what sort of political, social, and economic forces motivate, shape, and direct the research, development, and diffusion of advanced technologies and how do science and technology heuristically act to reshape or affect these forces.[9] Basic scientific research and applied technology are clearly important elements determining their own advancements, but various economic, political, institutional, social, and perhaps even individual psychological forces also influence these processes. The intellectual and policy implications of this assertion are obvious. Before it is possible to control the influence of a given technology on a society, it is critical to understand what forces shape that technology's development and application and how. From a policy perspective, the issues are how a technology can be directed and what are the implications of government intervention.

Economy Theory versus Political Practice," *Research Policy* 3, 1 (April 1974):3-16.

[6] Edwin Mansfield has written at great length on the relationship of government and its sponsorship of technical advancement; see *The Economics of Technological Change* (New York: Norton, 1968).

[7] George Eads and Richard R. Nelson, "Government Support of Advanced Civilian Technology: Power Reactors and the Supersonic Transport," *Public Policy* 19, 3 (Summer 1971):405-427.

[8] See Richard R. Nelson, Merton J. Peck, and Edward D. Kalachek, "Public Policy Issues," *Technology, Economic Growth, and Public Policy* (Washington, D.C.: The Brookings Institution, 1967).

[9] A literate set of examples is provided by Elting E. Morison, *Men, Machines, and Modern Times* (Cambridge, Mass.: The MIT Press, 1966).

A great deal of research has been performed on the first two of these broad policy questions. The third, however, is still generally neglected except in specific case studies.[10] It is to this last set of policy questions and issues that this work is directed, though across a wider set of development experiences than an individual case study.

The Development of the Nuclear Power Reactor

The nuclear power reactor represents a technology whose development and dissemination were characterized by major government research, development, and later operating subsidies. Most of the research devoted to this topic has examined the American development and commercialization of the atomic power reactor for three principal reasons.[11] First, the U.S. experience is the most thoroughly documented. Second, the U.S. nuclear industry has constructed more nuclear reactors (as measured in megawatts of electricity or simple number of reactors built) than any other nation. And third, the history of the American development and commercialization of the light water nuclear power reactor is inherently interesting, both as history and didactically for policymakers faced with supporting and later disseminating civilian technological advances in areas in which government research and development (direct or sponsored) has been dominant. Given the burgeoning policy analytic community in the United States, it is not surprising that the development of the U.S. nuclear reactor system has received extensive attention.

---

[10] The Apollo moon landing and the development of the multiple missile warhead are two examples where institutional and national politics had a major influence on technical progress. See John M. Logsdon, *The Decision to Go to the Moon: Project Apollo and the National Interest* (Cambridge, Mass: The MIT Press, 1970); and Ted Greenwood, *Making the MIRV: A Study of Defense Decision Making* (Cambridge, Mass: Ballinger, 1975).

[11] Robert L. Perry et al., *Development and Commercialization of the Light Water Reactor, 1946-1976* (Santa Monica, Calif.: The Rand Corporation, R-2180-NSF, August 1977), is a comprehensive overview of the U.S. light water nuclear power reactor program and the attendant industry; hereafter referenced as Perry et al., *The Light Water Reactor*. Frank G. Dawson, *Nuclear Power Development and Management of a Technology* (Seattle: University of Washington Press, 1976), offers a more technical perspective.

These reasons are undoubtedly true and relevant, but they overlook two broader issues. First, the developmental experience regarding atomic reactors is not unique to the United States, even given American preeminence. As *Scientific American* recently pointed out, "In the midst of the current debate over the appropriate role of nuclear power in the future U.S. energy budget it is easy to lose sight of the fact that elsewhere in the world the commitment of nations to the generation of electricity from atomic fission is proceeding apace."[12] Bupp and Derian note that over 60 percent of the world's operating and planned nuclear power plants are outside the United States.[13] The United States, which produced approximately "8 percent of its electricity from nuclear power, ranks behind . . . Switzerland (18 percent), West Germany (15 percent), Sweden (13 percent), the United Kingdom (10 percent), and France (10 percent)."[14] If one wishes to examine the R&D history of the nuclear power reactor and its diffusion for policy-relevent ramifications that might be applicable to the future direction or support of other high technology advancements (especially to other nuclear energy technologies), there is a much larger sample of pertinent experiences upon which to base the analysis.

Second, focusing on the U.S. experience with nuclear power reactors might lead one to overlook the more general area of government development of expensive, complex, and advanced technologies within a society. The examination of the research, development, demonstration (RD&D), and diffusion processes of nuclear power reactors in the United States, Great Britain, the Soviet Union, France, Canada, and the Federal Republic of Germany would provide a sufficiently varied set of cases and experiences that one might begin to construct a general model of

---

[12] "Nuclear Power Abroad," *Scientific American* 235, 2 (August 1976):44A.

[13] Irvin C. Bupp and Jean-Claude Derian, "The Nuclear Power Industry," in *The Management of Global Issues*, Appendix B of *The Commission on the Organization of the Government for the Conduct of Foreign Policy* vol. 1, (Washington, D.C.: GPO, 1975), p. 86. Also see Carl Walske, "Nuclear Electric Power and Proliferation of Nuclear Weapon States," *International Security* 1, 3 (Winter 1977): pp. 96-97.

[14] See "Nuclear Power Abroad." Walske, ibid., cites the same figures but includes Belgium (10 percent) and credits the United States with generating 9 percent of its electricity from nuclear power stations.

advanced technology development and application. Such a model should encompass the political, social, and economic aspects of technology development and diffusion as well as the more standard technological patterns of development.

There are three reasons for carefully examining and detailing a number of comparable experiences of national developments of nuclear power reactors. First, they have an intrinsic interest as histories and as comparative policy analyses. No single published source analyzes and contrasts the development of more than a few nations' reactor development and dissemination programs. Furthermore, the comparative aspect might illuminate different cultural variables as they affect technology. Second, these developments might provide insights into a set of immediate policy issues—the development and dissemination of new energy resources, particularly those dealing with nuclear or thermonuclear forms of energy. The historical interaction between civilian nuclear reactor programs and military weapons programs could add valuable evidence to the currently debated proposition that peaceful nuclear research is the most expedient way of acquiring a nuclear weapons capability.[15] In addition, the internal structuring of several national markets might offer some insights into the future structuring of an international nuclear market.[16] Third, such studies could serve as the data base for deriving a general model of technology development and application.

This study is based on comparative analysis examining similarities and difference among the six national programs at various stages of the RD&D and diffusion process. It has three general approaches: technologies, political and research institutions and decisions, and economics. The first refers to the

---

15 These fears are described by Albert Wohlstetter, "Spreading the Bomb Without Quite Breaking the Rules," *Foreign Policy* 25 (Winter 1976-1977):88-97, 145-170. This article is part of a more detailed ACDA-sponsored study, Albert Wohlstetter et al., *Moving Toward Life in a Nuclear Armed Crowd?* (Los Angeles: Pan Heuristics, ACDA/PAB-263, April 1976). A more optimistic assessment is offered by Walske, "Nuclear Electric Power."

16 See Abraham A. Ribicoff, "A Market-Sharing Approach to the Nuclear Sales Problems," *Foreign Affairs* 54, 4 (July 1976):763-787; Paul L. Joskow, "The International Nuclear Industry Today," *Foreign Affairs* 54, 4 (July 1976):788-803; or Steven J. Baker, "Nuclear Proliferation: Monopoly or Cartel?" *Foreign Policy*, 23 (Summer 1976):202-220.

technology requirements and developments that generated the nuclear power reactor and brought it to fruition. The second refers to the organizations responsible for the RD&D of the reactor technologies and their political environment.[17] The third refers to the economic forces that influenced many of the pertinent decisions. All three closely interact with one another, each at times motivating, leading, or possibly obstructing the others.

Comparative Policy Analysis

These research objectives require dealing with the problems inherent in any cross-national or comparative policy analysis. Although the comparison of national institutions, political structures, cultures, and economies is central to many academic disciplines, the cross-national comparison of national technologies and R&D strategies is fairly inchoate.[18] The comparative, interdisciplinary analysis of national experiences for policy purposes is even less developed and has only recently begun to receive attention.[19] The reasons are not difficult to discern.[20] Comparative policy studies, by definition, are broad, re-

---

[17] Giandemenico Majone, "Technology Assessment and Policy Analysis," *Policy Sciences* 8, 2 (June 1977):175, paraphrases Kant: "Technology Assessment without institutional analysis is empty; institutional analysis without technology assessment is blind."

[18] Some exceptions are found in Rand research on national weapons systems acquisition strategies: Robert Perry, *Comparisons of Soviet and U.S. Technology* (Santa Monica, Calif.: The Rand Corporation, R-827-PR, June 1973), and Robert Perry, Giles K. Smith, Alvin J. Harman, and Susan Henrichsen, *Systems Acquisition Strategies* (Santa Monica, Calif.: The Rand Corporation, R-733-PR/ARPA, June 1971). Specific comparisons are made of national R&D strategies by Arthur J. Alexander, "Weapons Acquisition in the Soviet Union, United States, and France," in Frank B. Horton, II, Anthony C. Rogerson, and Edward L. Warner, III (eds.), *Comparative Defense Policy* (Baltimore: The Johns Hopkins University Press, 1973), pp. 426-433.

[19] The American Political Science Association prize for the best work in the area of U.S. National policy published in 1975 was awarded to Arnold J. Heidenheimer, Hugh Heclo, and Carolyn Teich Adams, *Comparative Public Policy: The Politics of Social Choice in Europe and America* (New York: St. Martin's Press, 1975), a book on public policy that is explicitly comparative in its scope.

[20] See Arthur Cyr and Peter deLeon, "Comparative Policy Analysis," *Policy*

quire large amounts of scarce resources (both fiscal and personal), and the final products have not yet demonstrated their value and relevance. Even if funds and interests were coincident, the intellectual and analytical hurdles are formidable. The data are often dissimilar, difficult to obtain, or even nonexistent for a variety of reasons, such as national classifications for security purposes, the lack of substantive evidence, and the proprietary nature of some of the information. Comparative analysis could be difficult because different political, economic, and social structures and systems render the analytic comparison of various issue areas problematic at best, and misleading at worst. For example, a straightforward equating of Soviet R&D expenditures on a given technology with American expenditures on the same technology would be convenient but erroneous; Soviet bundles of goods are purchased and utilized in different ways than their counterparts in American society.[21] Finally, the "uniqueness" of the different conditions and situations vitiates easy judgments about possible patterns or trends. The implication is that some patterns that seem to be critical must remain qualified until additional research can validate their existence and importance.

All of these problems are relevant to the comparative analysis of national R&D programs leading to nuclear power reactors and their later dissemination. Yet important advantages can be gained from this multinational approach. Simply put, to evaluate different approaches to a problem, one must examine a number of different approaches. The six nations' developmental experiences detailed here share noteworthy common phenomena, so dissimilar processes are not being compared. The investigation of multiple cases of similar experiences provides the analyst with a wider data base from which to draw policy implications. The larger sample alleviates the potential hazards endemic to a single,

---

Sciences 6, 4 (December 1975):375-384, for a more thorough discussion of the origins, problems, and potential advantages of comparative policy analysis.

[21] It has only recently been acknowledged that Soviet and U.S. defense expenditures cannot be equated on a dollar-to-dollar (or ruble-to-ruble) basis. See Les Aspen, "How to Look at the Soviet-American Balance," *Foreign Policy* 22 (Spring 1976):96-106. Also see the testimony of Andrew W. Marshall before the Joint Economic Committee, reprinted as "Estimating Soviet Defense Spending," *Survival* 18, 2 (March/April 1976):73-79.

strictly historical narrative and reduces the possibilities of drawing critical but potentially misleading policy conclusions from a single idiosyncratic case. At the same time, this approach permits the identification and discussion of discrepancies and anomalies that would have to be recognized and included in any attempt to formulate a general theory of technology development. The necessary reservations against unwarranted interpretation and extrapolation will be noted where relevant. Still, notwithstanding such caveats, useful observations can be made regarding government programs of technology development and proliferation.

Scope and Focus

The scope of the case studies that are to be compared here is restricted to the research, development, demonstration, and diffusion of "slow" nuclear fission reactors (such as the light water reactor and the graphite-gas reactor)[22] whose primary function is the generation of electricity. This implicitly includes research reactors that were developed as part of the RD&D process leading to power reactors and excludes the numerous national RD&D programs for "fast" fission reactors (e.g., the liquid metal fast breeder reactor), fusion or thermonuclear reactors, and nuclear weapons systems. Obviously the development of nuclear energy sources cannot be viewed in pristine isolation from other nuclear research projects. The U.S. commercial light water reactor was partially an outgrowth of the U.S. Navy's nuclear submarine propulsion program,[23] and each nation included here has considered or actually undertaken a fast breeder reactor development program as a long-range energy alternative in view of the limited and diminishing amounts of uranium.[24] Other nuclear research programs will be referred to only when they are integrally related to those explications.

---

[22] The different reactor technologies are explained in Appendix B.

[23] See Richard G. Hewlett and Francis Duncan, *Nuclear Navy 1946-1962* (Chicago: University of Chicago Press, 1974).

[24] For recent descriptions of the French and American breeder reactor programs, respectively, see Georges A. Vendryes, "Superphenix: A Full-Scale Breeder

A general shortcoming in much of the earlier comparative literature is the tendency to consider and treat nations as singular or unitary decisionmaking bodies. Nations are anthropomorphized as rational actors or economic men. This level of analysis falls into what Allison has termed the National Actor, the Model I, paradigm.[25] A second level of analysis focuses on the organization, its subunits, and its personnel. It treats them as competitive entities, contending with other organizations for the right to determine, or at least influence, policies and actions that they consider as falling into their institutional domain.[26] This level is a combination of Allison's Models II and III.[27] A third level of analysis, perhaps derived from microeconomics, treats the individual as the unit of analysis.[28] A subset of this level examines the individual's role as a decisionmaker. Within recent years, a literature has been developed that explicitly documents the influence of individuals on public policy with the aid of psychoanalytic insights.[29] Steinbruner has combined the Neustadt-Allison "bureaucratic politics" models with the research of social psychologists into what he calls a "cybernetic paradigm" of decisionmaking.[30]

Reactor," *Scientific American* 236, 3 (March 1977):26-35; and Leland L. Johnson et al., *Alternative Institutional Arrangements for Developing and Commercializing Breeder Reactor Technology* (Santa Monica, Calif.: The Rand Corporation, R-2069-NSF, November 1976).

25 Graham T. Allison, *Essence of Decision: Explaining the Cuban Missile Crisis* (Boston: Little, Brown, 1971), ch. 1.

26 "Domain" in this sense follows James T. Thompson, *Organizations in Action* (New York: McGraw-Hill, 1967)—the political or functional areas in which an institution has recognized (although not necessarily undisputed) jurisdiction.

27 Allison, *Essence of Decision*, chs. 3 and 5.

28 Douglas E. Ashford, "Political Science and Policy Studies: Towards a Structural Solution," *Policy Studies Journal* 5, Special Issue (June 1977): 575-576.

29 For example, see Bruce Mazlich, *Kissinger: The European Mind in American Policy* (New York: Basic Books, 1976); or Alexander L. George and Juliette L. George, *Woodrow Wilson and Colonel House* (New York: John Day, 1956), for two examples of this approach. Robert C. Tucker, "The Georges' Wilson Reexamined," *American Political Science Review* 71, 2 (June 1977):608-618, offers a thoughtful evaluation of psychobiography.

30 John D. Steinbruner, *The Cybernetic Theory of Decision* (Princeton: Princeton University Press, 1974), especially Part I. The policy application of the cybernetic model are tested by Robert F. Coulam, *Illusions of Choice: The F-111 and the Problem of Weapons Acquisiton Reform* (Princeton: Princeton University Press, 1977).

This book will primarily rely upon the organizational level of analysis. The National Actor model, although unquestionably convenient, does not reflect the richness and complexity of the developmental and decisionmaking processes, nor is it sufficiently discriminating to permit making policy relevant observations. The amount of data collection necessary to operate at the individual level of analysis is monumental and, given the scope of the analysis, would almost certainly be infeasible. Even if the information were obtainable, to identify the pertinent patterns upon which to base policy conclusions from the mass of evidence would be virtually impossible. This level of analysis is too detailed, too rich with information, and would undermine the comparative aspect of this work. For these reasons, the level of analytic attention for this book will be the organization, as well as how the various participating organizations interacted with each other and the differing technologies. As Majone notes, "In assessing technology, we are really evaluating institutions."[31]

The dimensions of time for the various national programs considered here are not synchronous. Not all national programs were initiated at the same time. Canada and the United States completed their fundamental nuclear R&D by the mid-1960s, chose a reactor type, and all that remained was the general acceptance and dissemination of the reactor for power generation by electric utilities, which in these cases followed in an orderly fashion. France and the United Kingdom continued to make important alterations in their nuclear reactor programs throughout the 1960s and into the early 1970s. Thus, the analysis is less of a chronological or historical equating of the national programs and more a functional, programmatic comparison, even though the pace of the relative developments is itself a subject of interest and will be addressed.

Finally, none of these programs may be viewed in isolation from the others, even though each seems to progress of its own accord or by its own momentum. Each developmental program was, in some degree, responding to exogenous pressures in the political, economic, and technological senses of the term. This is most readily apparent for France, which clearly initiated its

---

[31] Majone, "Technology Assessment and Policy Analysis," p. 175.

nuclear research programs and later the development of nuclear weapons to reestablish French hegemony in Europe and to assert its independence from the United States.[32] However, each of the nations involved weighed its domestic nuclear power reactor technology against those of other nations, and in some cases, its decisionmakers considered adopting another nation's nuclear technology rather than develop its own. Thus, the nations' nuclear RD&D cannot be viewed as independent progressions or mutually exclusive.

Organization

This book is divided into three parts. Part I, "Advanced Technology and Public Policy," contains two chapters, which introduce the issues to be examined, the research and policy rationale, and the definition of the research effort, and set out in some detail the attributes and objectives that define a technology development and diffusion in the context of an applied analytic framework.

Part II, "The Comparative Analysis of National Reactor Programs," examined the RD&D and diffusion of nuclear reactors for six nations. Chapter 3 discusses the initial decision in these nations to develop a nuclear research program, and Chapter 4 identifies the early participants in these programs. Chapter 5 compares the national decisions to develop either single or multiple reactor options. Chapter 6 describes the reactor demonstration phase, and Chapter 7 relates the succeeding dissemination decisions.

Part III, "Conclusions," reviews the previous part in light of the proposed framework and the possible policy conclusions. Chapter 8 suggests the policy conclusions derivable from the national developmental experiences that appear to be generally applicable to public policy issues of technology development.

---

[32] See Bertrand Goldschmidt, *The Atomic Adventure: Its Political and Technical Aspects*, translated by Peter Beer (New York: Macmillan, 1964); or Lawrence Scheinman, *Atomic Energy Policy in France Under the Fourth Republic* (Princeton: Princeton University Press, 1965).

Chapter 9 places the reactor development experiences in the context of other technology developments and proposes research that lays the foundation for a more general model for understanding technology development.

2

An Analytic Framework for Comparing Government-Sponsored Civilian Technology Developments

The necessity for an analytic framework to structure and guide the analysis is critical to any research effort, particularly when the evidence is drawn from cross-national case studies of public policy behavior in which the data are often quite disparate.[1] Such a framework must be sensitive to unique national characteristics or attributes affecting policy behavior while providing a means for identifying common attributes and behavior patterns.[2] Unless the characteristics of various national programs can be formulated into a framework so that patterns of institutional rela-

[1] This deficiency is apparent in the literature of comparative politics, even in the Social Science Research Council's seventeen-year program to delineate a model of political development (as set forth in the Princeton University Press' Series on Political Development). For a careful analysis of this particular shortcoming in the Political Development Series, see Robert T. Holt and John E. Turner, "Crises and Sequences in Collective Theory Development," *American Political Science Review* 69, 3 (September 1975):979-994.

[2] See Sidney Verba, "Some Dilemmas in Comparative Research," *World Politics* 20, 1 (October 1976):111-127.

tionships applicable across the sample may be discerned, the case studies and data would be quite resistant to any policy-oriented research.[3] The purpose of such a framework for this exercise, then, is twofold. First, it should permit the structuring of a large amount of material on six national nuclear power reactor development and dissemination programs so that policy-relevant observations can be discussed. Second, it should be useful in examining other programs of technology development that fit the broad criteria delineated below and, at the same time, suggest the basis for a general model of organizations or institutions as they influence technology development and the converse.

To these ends, a tentative analytic framework is proposed here for the RD&D and diffusion of technology developments that share the following general characteristics:

1. *Significant amounts of government subsidy and control.* A number of possible reasons for dominant government intervention can be posited: for example, high R&D expenditures that seemingly price the proposed development beyond the financial capability of private industry, the recognition that the end product would be a "public good" and therefore warrants public funding, and the potential hazards of uncontrolled development and diffusion. For whatever reason, the model of technology development set forth below specifically applies to developments that manifest a significant degree of government sponsorship, intervention, and direction.
2. *Technology destined for direct civil consumption* or the civilian marketplace. This explicitly precludes strictly military weapons systems developments or projects lacking tangible civilian consumption (e.g., Apollo).
3. *Technology monitored from the basic R&D phase through the demonstration phase into the diffusion or dissemination period.* In this characteristic, the framework differs from earlier works that were primarily interested in the basic R&D,

---

[3] See Cyr and deLeon, "Comparative Policy Analysis," p. 376.

product innovation periods, or the commercial demonstration and dissemination phase.[4]

4. *The development, demonstration, and diffusion of unproven technologies*. The construct is not meant to address the development and diffusion of proven technological or managerial innovations. A high degree of technological uncertainty must be present.

These stipulations exclude a great many programs, especially military. Although the experience of some of these programs are drawn upon in formulating the general framework, the thrust of the analysis is civilian. The exclusion of military programs does not preclude the use of suitable cases. Aside from nuclear power reactors, several technology development programs fit easily within these defining characteristics in such diverse fields as civil aviation and maritime systems (the SST and nuclear ship "Savannah"), rapid transit (particularly personal rapid transit systems), communications satellites, medical research (e.g., computer-assisted electrocardiogram analysis and hydraulic knee prosthetic devices), and saltwater conversion plants.[5] Future developments, such as new energy technologies, would also be included within the scope of the analytic structure.

It is now possible to describe a general framework for analyzing technology developments falling within these criteria. After the case studies, this framework will be used to extract policy implications from national nuclear reactor development experiences. The framework and data will also be used to propose a general model of government-sponsored technology developments.

The Analytic Framework

The main concern of the analysis is the institutional drama that shaped and directed the development and dissemination of ad-

[4] For an example of the latter, see Walter S. Baer, Leland L. Johnson, and Edward W. Merrow, "Government-Sponsored Demonstrations of New Technology," *Science* 196, 4293 (May 27, 1977):950-957.

[5] These specific examples (excluding communication satellites and the SST) are

vanced technologies, specifically nuclear power reactors. This analysis refers to the organization or institution itself as the unit of analysis and does not further break it down to the suborganizational level or unravel its standard operating procedures. Even though the organization is, of course, an amalgam of component organizations, the interactions examined here are inter- rather than intraorganizational in context.[6] Instead of comparing specific institutions, such as the government agencies charged with the development of atomic energy (e.g., the American Atomic Energy Commission or the British Atomic Energy Authority), the analytic framework will be much more generic so that the structure may later be used for other types of technology developments beyond scope of this book.

The proposed analytic framework for technology development and dissemination is predicated on a series of interactions between a set of institutional *objectives* that represent various political and economic goals underlying the various technology developments and a set of *attributes* that affect and characterize these same developments. The relative importance of the posited objectives and attributes should change during the course of the development as the political, social, and economic contexts change.

The key consideration for the sets of objectives and attributes is not their exactitude in structuring a rigorous, predictive framework.[7] More appropriate criteria of the proposed lists are their utility, inclusiveness, and internal consistency as they are defined, rather than their premature precision.[8] The first criterion asks if the defining sets of objectives and attributes are robust enough to explain the observed phenomena while being sparse enough that one is not overwhelmed with multiple categories and conditions. The second condition is that the sets are sufficiently

---

drawn from and detailed in Walter Baer et al., *Analysis of Federally Funded Demonstration Projects: Supporting Case Studies* (Santa Monica, Calif.: The Rand Corporation, R-1927-DOC, April 1976).

6 See James G. March, "The Business Firm as a Political Coalition." *Journal of Politics* 24, 4 (November 1962):662-678.

7 See Abraham Kaplan, *The Conduct of Inquiry: Methodology for Behavioral Science* (San Francisco: Chandler, 1964), Section VII, "Models."

8 Problems with identifying and obtaining possible measures for each of the objectives and attributes are discussed in Appendix A.

explanatory to encompass the category of technology developments covered here. The third criterion demands that the sets observe the same phenomena and posit a different set of motivations and explanations.

Objectives

Earlier policy works on technology development within a single nation have been somewhat restricted in their intentions and have therefore been able to define a quite discrete and singular criterion or objective for measuring the "success" of a given program. For example, Baer, Johnson, and Merrow in a study of federally funded technology demonstration programs, suggest that the government agency's objective in sponsoring these demonstration projects was limited to addressing the financial and technological uncertainties that, if favorably resolved, would result in the commercial (in most cases, federally unsubsidized) acceptance and dissemination of the technology.[9] However, a full appreciation of the technology development phenomenon is a more complex undertaking. Government sponsorship of the RD&D processes brings with it a number of participating organizations,[10] each with its own set of objectives, such that the definition of a single objective and later evaluation of success or failure of a development as a function of that objective is exceedingly difficult.[11] Even in private industry, with the ubiquitous dollar criterion, singular objectives are not always identifiable. The loss of over $300 million was ample indication to the Ford Motor Company that the Edsel automobile should be discontinued within a few years of its widely heralded introduc-

---

[9] Baer, Johnson, and Merrow, "Government-Sponsored Demonstrations of Technologies."

[10] See George D. Greenberg et al., "Developing Public Policy Theory: Perspectives from Empirical Research," *American Political Science Review* 71, 4 (December 1977):1532-1543.

[11] There is, of course, a substantial literature on organizational objectives and goals; for a concise review, see Lawrence B. Mohr, "The Concept of Organizational Goal," *American Political Science Review* 67, 2 (June 1973): 470-481.

tion,[12] but certainly numerous counterexamples may be found in which a company was willing to incur long-range losses for reasons of prestige, sentiment, or the hopes that the losses would later be turned into profitable enterprises. For example, IBM decided to enter the photostatic copier market against Xerox, which at the time held close to 80 percent of the market. Technology developments usually reflect multiple objectives, particularly when the public sector is involved.

The few attempts to compare national technology development programs have used a fairly simple, single-measure objective. For example, Costello and Hughes compared the British/French development of the SST with the analogous American and Soviet developments.[13] Their implicit statement of the national objective was the production of the first supersonic transport to fly paying passengers on a regularly scheduled route. By this measure, the French/British SST was more "successful" than its American and Soviet counterparts.[14] Joskow argued that the sole criterion for judging a national nuclear power reactor RD&D program was to produce a viable reactor system.[15] Utilizing this measure, he concluded that the key variable distinguishing the "successful" West German program from the less "successful" British and French programs was the presence of an actively involved private industry. Burn employed a similar singular objective to differentiate between the "successful" American light water reactor (LWR) development and the "failure" of the British reactor program.[16]

However convenient this assumption might be, it is

---

12 John Brooks, *The Fate of the Edsel and Other Business Adventures* (New York: Harper & Row, 1959).

13 John Costello and Terry Hughes, *The Concorde Conspiracy* (New York: Charles Scribner's Sons, 1976); a more critical assessment is in Andrew Wilson, *The Concorde Fiasco* (Baltimore: Penguin Books, 1973).

14 In their last chapter, Costello and Hughes argue that U.S. policymakers were ill-advised in their decision to abandon the SST. That opinion is, of course, open to debate.

15 Paul L. Joskow, "Research and Development Stretegies for Nuclear Power in the United Kingdom, France, and Germany," in *Proceedings of the Workshop on Institutional Alternatives for LMFBR Development and Commercialization*, (McLean, Va.: The MITRE Corporation, 1977), pp. 227-328.

16 Duncan Burn, *The Political Economy of Nuclear Energy* (London: Institute of Economic Affairs, 1967).

misleadingly simplistic in the context of national government programs on the scale considered here. Large organizations, their decisions, and the resulting policies and programs have been shown to reflect multiple, sometimes competing, objectives.[17] One well-known example of divergent behaviors within large institutions is Allison's description of the multiple objectives held by different organizations within the Soviet Union leading to the installation of medium-range ballistic missiles in Cuba in 1962.[18] Especially for international comparisons, to posit a singular objective would sorely distort the complexity of governmental decisionmaking and neglect a great deal of important information. As Verba has argued, "The patterns for which we try to account in comparative politics . . . are immensely complicated and cannot be accounted for by one or two simple causes. Only by considering the combination of a large number of factors can we hope to begin to account for relevant variations."[19] The importance of multiple objectives for national technology development efforts is emphasized by Greenwood's observation of the motivations behind the U.S. development of the multiple nuclear warhead:

> To adopt any of these practical explanations, the determination of technology, the inexorable drive of bureaucratic process or the preeminent role of central decisionmakers, as *the* explanation of the MIRV programs is to miss the richness and diversity of the decisionmaking process. All were involved and all must be included if an accurate explanation of the MIRV programs is to be given.[20]

In short, positing a single objective for a national RD&D program does not permit us to understand the full range of decisions and choices manifested by the multiple organizational actors who participate in and define national technology development programs.

Four possible objectives or reasons are posited why a nation's institutions might decide to develop and disseminate various forms of high level technology. This set of objectives explicitly

---

[17] The basic reference is Richard M. Cyert and James G. March, *A Behavioral Theory of the Firm* (Englewood Cliffs, N.J.: Prentice-Hall, 1963).

[18] Allison, *Essence of Decision* chs. 3 and 5.

[19] Verba, "Some Dilemmas in Comparative Research," p. 117.

[20] Greenwood, *Making the MIRV*, p. 81; emphasis added.

assumes that major national programs of technology development cannot and should not be measured against a single criterion.

The set of objectives includes:

Scientific Research for Science's Sake
National and International Prestige
Development of Efficient Technology
Political Equity

The various national decisionmakers who advocate (for whatever reasons) the development of a given technology obviously do not weigh each of these objectives equally; indeed, an individual decisionmaker probably does not even recognize or at least act upon all of these objectives, nor would they necessarily all be present at the same time with the same weight. However, these objectives exist in most national endeavors to develop given technologies and help explain the actions of these national actors with a richness unmatched by any single explanatory criterion.

Scientific Research for Science's Sake. Research in "pure" science differs from applied scientific or engineering research in that the practical applications of the research findings are not the central issue; the purpose is advancement of science itself rather than the immediate benefits to the society, and the research can be viewed as a "public" good. Example of scientific research as a public consumption good would include support for high energy physics and biological research; project MOHOLE and high atomic particle accelerators would be impossible without extensive government support.[21] Since the end of World War II, such research has been funded by the National Science Foundation, the Office of Naval Research, the Atomic Energy Commission, and the National Institutes of Health.[22] Initially, the principal

[21] See Daniel S. Greenberg, *The Politics of Pure Science* (New York: New American Library, 1967), Book III, for a narrative of these two programs.

[22] W. Henry Lambright, *Governing Science and Technology* (New York: Oxford University Press, 1976), especially ch. 6, describes the work of these agencies; also see Greenberg, *The Politics of Pure Science*, passim.

beneficiaries of this objective would be a nation's scientific personnel, who would directly profit in terms of research support, scientific discovery, and peer recognition.[23]

The objective of *Science for Science's Sake* addresses the amorphous goals of scientific advancement and, more distinctly, directly satisfies the research interests of a limited number of research personnel within the scientific communities and their sponsors and sympathizers in the government. This objective acknowledges the OECD statement that during the 1960s, "fundamental research became simply one national objective among others."[24]

National and International Prestige. One reason for committing large amounts of government resources to the development of a technology could be prestige, both (or either) international and domestic. Although the Apollo effort does not fall into our definition of a civilian technology development, President Kennedy's main consideration in dedicating that massive application of technology to the pursuit of a national goal was the desire to garner the immense international prestige that would accrue to the first nation landing a man on the moon.[25] Kennedy was less committed to a lunar landing than he was to any scientific triumph that would regain for the United States the acclaim the Soviets achieved with their first man in space. Kennedy was quoted at the time: "If you had a scientific spectacular on this earth that would be more useful—say desalting the ocean—or something just as dramatic and convincing as space, we would do that."[26] Similar sentiments were voiced by the German cabinet

---

[23] Don K. Price, *The Scientific Estate* (Cambridge, Mass.: Harvard University Press, 1965), lists the scientific estate as one of his four groups that determine science and technology policy within the American political system.

[24] G. Caty et al., *The Research System: Comparative Survey of the Organisation and Financing of Fundamental Research* (Paris: Organisation for Economic Co-operation and Development, 1974), Book III, p. 167.

[25] The bureaucratic obstacles to Kennedy's decision and the general implications of such decisions for decisionmaking theory are discussed by Paul R. Schulman, "Nonincremental Policy Making: Notes Toward an Alternative Paradigm," *American Political Science Review* 69, 2 (December 1975):1354-1370.

[26] Logsdon, *The Decision to Go to the Moon*, pp. 110-111.

minister announcing the government's decision to build nuclear power reactors and by French and British spokesmen in support of their national SST commitment. The British Supersonic Transport committee was explicit to this point:

> Since this country's future will depend on the quality of its technological products and since its scientific manpower and resources are less than those of the U.S.A. and U.S.S.R., it is important that a reasonable proportion of such resources are deployed on products which maintain our technical reputation at a high level. A successful supersonic aircraft would not only be a commercial venture of high promise but would also be of immense value to this country as an indication of our technical skills.[27]

The prestige motive could be for domestic consumption. A great deal of domestic support existed for the American SST as a symbol of American technology in general and the excellence of American aviation in particular. Similar sentiment was expressed by some public utility executives in the United States to the effect that nuclear energy was the "way of the future" in which they wished to participate.[28]

Prestige might also be international in intent. A nation's R&D establishments as well as the users might argue that it was necessary to be in the technological vanguard of the international scientific community. The president of American Airlines was quoted in late 1960 following an IATA conference called to discuss the potential of an American SST: "One of the worst beatings we'll ever get is if we have to look up to a Soviet SST, the way we had to look up to the first Sputnik."[29] Only one day after Pan American announced its decision to purchase options on at least six Concordes, President Kennedy committed the United States "to spending billions of dollars on a high-risk project to *satisfy national pride.*"[30]

Development of Efficient Technology. The assessment of this objective will obviously vary from technology to technology but its

---

27 Quoted in Costello and Hughes *The Concorde Conspiracy*, pp. 27-28.

28 John T. Hogerton, "The Arrival of Nuclear Power," *Scientific American* 218, (February 1968):29.

29 Quoted in Costello and Hughes, *The Concorde Conspiracy*, p. 38.

30 Ibid, p. 65; emphases added.

general intent is to provide a measure of the ability of a nation's R&D program to deliver a new technology capable of competing on a price basis with the other technology alternatives available to that nation for fulfilling a specified demand. In other words, can a nation's RD&D processes deliver a nuclear reactor capable of producing electricity at cost competitive with other methods of generating electricity (such as coal or oil burning) or a personal rapid transit (PRT) system able to compete with private automobile or municipal bus lines?

The problem of government subsidies interferes with a clean interpretation of technical efficiency or economic viability. For instance, a federal government grant might subsidize the capital construction costs of a PRT system for a city; its operations and maintenance expenses would be paid for out of passenger fares, but in all likelihood these expenses would be underwritten by local taxes.[31] Therefore, the operation of a PRT system might not constitute a "fair" test of its acceptance. However, there are ways of approximating the end product of the delivery system. For rapid transit, the construction and equipping of the Bay Area Rapid Transit (BART) system has cost a great deal more than originally expected and has operated at reduced schedules with heavy losses since its opening day.[32] The latter suggests that an efficient use of rapid transit technology has not been achieved. For nuclear reactors, the attainment of this objective might be more clearcut as the costs of electricity generated by commercial-sized nuclear reactors are compared with costs of other forms of electricity production.

Another reason for pursuing an efficient technology would be the explicit development of an export commodity that could exploit an international market, thus benefiting certain sectors of a nation's industry and, simultaneously, the national balance of trade. If a competitive technology could be developed, in all likelihood it could be exported to nations that wished to utilize that technology but did not have the resources to initiate their own RD&D processes. This condition can be observed at some

---

[31] See Walter Y. Oi, "The Federal Subsidy of Conventional Mass Transit," *Policy Analysis* 1, 4 (Fall 1975):613-658.

[32] Melvin M. Webber, "The BART Experience—What Have We Learned?" *The Public Interest* 45 (Fall 1976):79-108.

point in all of the nations that developed their own nuclear reactor technology with the exception of the Soviet Union. In that case, numerous reactors have been ordered by Communist bloc nations, but clearly these are not examples of reactor systems being purchased on the open market; only one Soviet reactor has been exported to a nonsatellite nation, India.[33]

The goal of developing an efficient technology for export purposes is quite clear. Government support of a given R&D program might produce an important technology that could be a major asset in a nation's international balance of trade calculations. For a nation whose trade balance appears precarious, such as Britain's after World War II, the development of an efficient, competitive technology for export purposes could be viewed as a particularly attractive investment and a worthwhile objective.

Political Equity. This objective represents the political process of resource allocation or reallocation within an economy. A government may choose to sponsor a particular technology development program as a response to demands made on the system by a number of political interest groups. It represents the resource distribution aspects of Lowi's public policy taxonomy.[34] Although economic theory describes a fairly mechanistic process moving throughout the economy toward a Pareto optimal condition,[35] this objective is specifically concerned with the allocation mechanism activated by political interest groups.[36] For example, government support for a particular technology development program may result from sectoral unemployment problems, as some have suggested is a partial rationale for the government

---

33 Gloria Duffy, "Soviet Export Policy," *International Security* 3, 3 (Winter 1978), offers an excellent discussion of the evolution of this policy.

34 Theodore J. Lowi, "Decision Making vs. Policy Making: Towards an Antidote for Technocracy," or "Four Systems of Policy, Politics, and Choice," in *Public Administration Review* 30, 3 (May/June 1970):314-325, and 32, (July/August 1972):298-310.

35 For a brief description of the principles underlying Pareto optimality, see Walter Nicholson, *Microeconomic Theory: Basic Principles and Extensions* (Hinsdale, Ill.: The Dryden Press, 1972), ch. 22. No criterion of efficiency is implied in the political equity process.

36 David B. Truman, *The Governmental Process* (New York: Alfred Knopf, 1951), is one of the basic works in group theory.

sponsorship of Boeing's development of the supersonic transport.[37] This was very much a rationale underlying the British development of the Concorde. Duncan Sandys, the Minister of Aviation in the Macmillan cabinet in 1959, flatly asserted: "If we are not in the supersonic aircraft business, then it's really only a matter of time before the whole British aircraft industry packs in. It's obviously the thing of the future. It may pay. It may not pay. But we cannot afford to stay out."[38] The influence of the business community on government policies has long been asserted, although it is rarely subjected to rigorous validation.[39] Other interest groups include labor unions and professional associations. The British coal mining unions were instrumental in slowing down the government's development of atom-powered electricity generating plants for fear that it would undermine the coal industry and the livelihood of their membership the political activities of scientists to secure government support for basic research were equally self-serving.[40]

The political equity objective is the allocation of economic resources among various interest groups in a politically or equitable or acceptable manner and reflects both the political resources of different interest groups and the governments' responses to them.

Together, these objectives offer a coherent set of rationales that might explain why a nation's policymakers chose to conduct and continue a major technology development program, as well as indicate some of the major decisions made during that program's development.

---

[37] James R. Kurth, "Why We Buy the Weapons We Do," *Foreign Policy* 11 (Summer 1973):33-56, contends that the backlog of a company's existing orders is an important consideration in the government's selection of contractors for major weapons systems.

[38] Quoted in Costello and Hughes, *The Concorde Conspiracy*, p. 39.

[39] See Raymond Bauer, Ithiel de Sola Poole, and Lewis Dexter, *American Business and Public Policy* (New York: Atherton Press, 1962); some empirical evidence is presented by Lester M. Salamon and John J. Siegfried, "Economic Power and Political Influence: The Impact of Industry Structure on Public Policy," *American Political Science Review* 71, 3 (September 1977):1026-1043.

[40] Greenberg, *The Politics of Pure Science*.

Attributes

A technology development conducted on a national scale can be characterized as affected by a number of "attributes" of the development system that were influential (by either their presence or their absence) in a large sample of technology developments. The same selection criteria and reservations that applied to development objectives apply here. The list of attributes should be quite inclusive, internally consistent (no duplication), general enough to capture a number of different phenomena, and specific (useful for discrimination among different technological programs and national programs). Such a list might include:

Active Scientific Community
Integrated Technology Delivery System
Concomitant Military Research Program
Early Starter
National R&D Heritage
Multiple Technology Options
Resources Invested

Active Scientific Community. This attribute is meant to capture the role of a nation's scientific community (or its patrons) in promoting a government's involvement in and financial support of large-scale technology developments. These scientists and technicians could be located in universities, national industries, or within the government itself. This attribute suggests more than participation by the scientific and technical communities; it implies the active sponsorship and an attempt at establishing priorities and procedures during the formative periods and later critical decisions. This activity could be pro or con (or both, as was the case with the SST controversy in the United States) and effective or ineffective. Examples of scientific activists include Walter Panofsky (and his advocacy of the Stanford Linear Accelerator) and Mary Lasker (in the National Institutes of

Health).[41] A representative question that might be asked is whether or not there are recognized spokespersons for the scientific community who are deliberately influential in governmental policymaking circles.

Integrated Technology Delivery System. This attribute describes the degree of cooperation or coordination among the major institutional actors in the development and dissemination of a technology. Wenk has hypothesized a "technology delivery system" (TDS) composed of (a) system inputs (e.g., knowledge, resources, personnel), (b) institutional structures, (c) system processes (how the institutions operate and interact), and (d) system outputs. Wenk employs TDS as a means of technology assessment.[42] His concept can be reformulated so that it focuses on the three principal institutional actors involved in the technology RD&D and dissemination processes: (1) the *developer*, usually a given agency of the central government, whose main function is to support basic scientific and R&D efforts; (2) the *vendor*, or the organization(s), whose primary responsibility is the development and assembly of the product to a point where a dissemination decision may be made; and (c) the *consumer* or purchaser of the new technology.[43] For a PRT, the TDS would be defined as (1) the developer within the Department of Transportation, especially the Urban Mass Transit Administration (UMTA), which funds PRT research and demonstration grants; (2) the vendor would be the aerospace industry, especially Boeing, Rohr, and Aerospace, which have conducted research and built demonstration facilities; and (3) the consumers, which would be various cities wishing modern rapid transit facilities, such as Denver, which

---

[41] Panofsky's advocacy is documented by Greenberg, *The Politics of Pure Science*; for a description of Lasker's activities in support of biomedical research, see Lambright, *Governing Science and Technology*, pp. 150-152.

[42] Edward Wenk, Jr., "Technology Assessment in Public Policy: A New Instrument for Social Management of Technology," *Proceedings of the Institute of Electrical and Electronics Engineers* 63, 3 (March 1975):378-379.

[43] W. Henry Lambright and Albert H. Teich, "Technology Transfers as a Problem in Interorganizational Relationships," *Administration and Society* 8, 1 (May 1976): p. 30, refer to "an administrative 'delivery system' " composed of an "innovator, a manufacturer, and a user."

was awarded a large UMTA grant to construct a PRT system.[44] The U.S. SST provides a second example. The Federal Aviation Agency was to have sponsored and funded the RD&D, Lockheed or Boeing was to have been the vendor, and the airlines would have been the consumer.

Three important reservations need to be observed. First, at present the existence of a TDS is largely hypothesized. The issue of actor integration in a social welfare system has been presented in terms of agency "cooptation"[45] and Baer and his colleagues have noted several instances of a TDS in much more limited government technology demonstration programs,[46] but the existence and effects of this relationship on very large and important government-sponsored research programs has not yet been carefully tested and delineated. Its existence in a multinational context is even more speculative; the two examples used above both come from a society in which there is a clear distinction between the government and private industry. One purpose of this research is to determine if a TDS and its component institutional actors can be identified in a variety of economic systems. Second, the ultimate consumer is, of course, the individual citizen, but for present purposes, we define the organization that accepts, purchases, and uses the new technology as the consumer. Third, the existence of a TDS does not imply a successful technology development; the systems that delivered the PRT and the Concorde have not been able to generate widespread acceptance of their labors.

The particular concern here is the degree of "integration" exhibited by a nation's TDS. This means the organizational independence or, more exactly, the degree to which the three actors

---

[44] See G. Christina Hill, "Troubled Transit: Some Experts Think 'Personal Transit' Is Wave of the Future," *Wall Street Journal* (August 30, 1974), pp. 1, 19. For a more pessimistic assessment of a PRT demonstration project, see Robert Lindsey, "U.S. May Raze Its $57-million Showcase 'People-Mover'," *New York Times* (April 13, 1974), pp. 1, 26.

[45] Philip Selznick, *TVA and the Grass Roots* (New York: Harper and Row, 1966).

[46] Walter S. Baer, Leland L. Johnson, and Edward W. Merrow, *Analysis of Federally Funded Demonstration Projects: Final Report* (Santa Monica, Calif.: The Rand Corporation, R-1926-DOC, April 1976), pp. 51-52; also Arthur A. Ezra, "Technology Utilization: Incentives and Solar Energy," *Science* 187, 4187 (February 28, 1975): pp. 707-713.

were able to coordinate their roles within and among the phases of the development and diffusion and the way in which the roles complemented or conflicted with each other. In short, the attribute being observed is the degree to which the TDS actors were cooperatively integrated in the national effort to produce a given technology. This suggests three distinct institutional roles (although obviously overlaps can and do occur) within which the actors have some degree of latitude. An integrated TDS also implies that no one actor or organization dominates the others throughout the research, development, and dissemination periods.

Concomitant Military Program. This attribute seeks to determine if a nation's civilian technology development programs were carried on in conjunction with a military program whose research could have positively benefited the civilian program. That is, did the civilian programs gain any "spillover" effects from a concomitant national military program? (Spin-offs from space technology research also could be considered here.) For example, then-Senator Lyndon Johnson argued that "Transportation of people and cargo at three times the speed of sound, and above the weather, could be attained through the utilization of B-70 technology,"[47] and many have suggested that the technology acquired by Boeing from its military bombers and tankers (the B-47, B-52, and KC-135 programs) acted as a government-funded prototype for the Boeing 707, the most successful civilian jet airliner built.[48] The initial U.S. nuclear power reactors were extensions of the nuclear propulsion technology that Westinghouse, the AEC, and the U.S. Navy developed for the Navy's nuclear submarine program.[49]

Early Starter. This attribute is meant to measure if the national technology development program was a pioneering effort (was it

---

[47] Quoted in Costello and Hughes, *The Concorde Conspiracy*, p. 33.

[48] The "widespread but mistaken belief that the federal government subsidized the Boeing 707 transport through the KC-135 tanker program" is discredited by Eads and Nelson, "Government Support of Advanced Civilian Technology," p. 408.

[49] Hewlett and Duncan, *Nuclear Navy*.

in the technological vanguard of specific technology development programs relative to other nations?) or was it an example of "buying in" at a later date when the technology had been largely developed elsewhere.[50] The latter behavior has often been the case in a great number of private industry developments (e.g., computer and electronic equipment). Gilpin has suggested that this choice might be a conscious decision in a nation's R&D strategy and has a significant effect on the program objectives.[51] It was particularly apparent for nuclear power development; only five nations have chosen to build their nuclear power reactor programs without significant investments in exogenous technology.

The other attributes are observed throughout the RD&D and diffusion processes. The measure here focuses on the development and dissemination periods as opposed to the early periods characterized by basic research. The question is whether the latter two periods were based largely upon indigenous research that roughly defined the technological frontiers of that technology (was that nation an "early starter"?) or were they based on proven, foreign designs?

R&D Heritage. This variable is meant to capture a nation's "track record" of past performance in large R&D or scientific projects. In other words, is there a history of technology developments and innovation diffusions? It also implies that presence or absence of the institutions and personnel making up that tradition. One effort to observe this attribute is the OECD attempt to specify preliminary patterns of basic research and development among its member states.[52] Other OECD studies have examined technological innovation and diffusion within its membership.[53] The critical assessment is the existence of

---

50 This, of course, opens the question of technology transfers. This framework does not consider the political implications of technology transfers or foreign licensing. It does acknowledge that such transfers exist and that their use may be applicable toward the fulfillment of the proposed objectives.

51 Robert Gilpin, "Technological Strategies and National Purpose," *Science* 169, 3944 (July 31, 1970):441-448.

52 See G. Caty et al., *The Research System*, vol. I (U.K., France, West Germany, 1972); vol. II (Belgium, Netherlands, Norway, Sweden, Switzerland, 1973), and vol. III (Canada, U.S., and General Conclusions, 1974).

53 For one attempt at systematically ordering such data, see OECD, *Gaps in*

previously productive R&D establishments and personnel upon which the specified technology development could be based.

Multiple Technology Options. The literature on weapons systems R&D relates that successful development projects requiring major technological advances are characterized by multiple and competitive research options.[54] Scherer asserts that "competition between substitutes affords the government two main benefits . . . the statistical benefits and the behavioral benefits,"[55] both of which ensure that the development does not invest all of its resources into a technology that ultimately is found to be inadequate. This approach is also applicable to government-funded civilian technology development. Translated into the proposed set of attributes, this variable can be viewed as considering the number of technological alternatives that a nation's R&D establishments brought as far forward in the RD&D process as the development phase or, even better, to the demonstration stage.

This is not meant to imply the ultimate worth of this approach to advanced technology development. The Canadians only developed and demonstrated one type of nuclear power reactor and seemed reasonably satisfied with its performance; UMTA has financed a number of different PRT developments and one demonstration project and found none of them satisfactory. This attribute is only meant to ask whether there were multiple technology options maintained sufficiently late in a technology development that the relevant policymakers or technology consumers had a choice in selecting among alternatives. For instance, U.S. electric utilities had a wide menu of reactor options from which they could choose, but U.S. airline companies were constrained to one SST model if they chose to buy American.

Technology: Comparisons Between Member Countries in Education, Research & Development, Technological Innovation (Paris: Organisation for Economic Cooperation and Development, 1970), Book III.

54 Thomas K. Glennan, Jr., "Issues in the Choice of Development Policies," in Thomas Marschak, Thomas K. Glennan, Jr., and Robert Summers, *Strategies for R&D: Studies in the Microeconomics of Development* (New York: Springer-Verlag, 1967), p. 47.

55 F.M. Scherer, *The Weapons Acquisition Process: Economic Incentives* (Cambridge, Mass.: Harvard University Press, 1964), p. 19.

Resources Investment. This attribute refers to the financial and human resources, both government and nongovernment, expended on a given technology development. It can be treated either as a cumulative expenditure or within developmental stages. This calculation should also include government subsidies for the purchase and operation (e.g., waiver of fuel costs) of a particular technology. It reflects that large amounts of money were spent over an extended time to develop a given technology. Resources invested are important to a development for two reasons. First, and most obvious, adequate resources must be invested in the project so that it can be carried to fruition. Second, and perhaps less obvious, the prior resources invested can have a significant effect on pending and future investment decisions.

The economic theory of the firm asserts that each investment or funding decision should be based on the marginal cost and the expected marginal return. Previous or "sunk" costs are already incurred and should not be confused with or substituted for marginal costs; only the latter should be considered when making investment calculations.[56] However, this theorem is patently normative and not descriptive of how sunk costs, in fact, influence future investment decisions, especially those with institutional or political components.[57] There are, of course, pressing organizational and political reasons beyond the ken of economic theory that explain why a government (or any sponsor) would deliberately reject marginal cost analysis and choose to honor sunk costs with continued funding. These reasons might include the fear of widespread unemployment, the loss of a critical industrial sector, bureaucratic intransigence, and a general unwillingness to admit a program's failures as signified by the termination of its funding. These examples stand in contradistinction to the "rational" calculus posited by the economic theory of the firm. The actual influence of sunk costs thus represents the economic and political baggage that previous investments in a technology bring to present decisions.[58] When faced with the growing technical obstacles

---

56 See Nicholson, *Microeconomic Theory*, Chapter 12.

57 This issue is addressed by Charles Wolf, Jr., "The Present Value of the Past," *Journal of Political Economy* 78, 4 (July/August 1970):783-792.

58 This "incremental" form of decisionmaking is illustrated by Aaron Wildavsky, *The Politics of the Budgetary Process*, 2nd ed. (Boston: Little, Brown, 1974).

and sharply rising costs presented by the Concorde, the French government was unwilling to abide by the rational economic dictum to ignore its sunk R&D costs and terminate the program. A similar rationale may be attributed to the British cabinet's decision to continue domestic research on the English nuclear reactor system despite a history of technical shortfalls and financial overruns.

This particular variable presents the most serious data and definitional problems in the analytic framework. Often it is impossible to identify sunk costs as a decisive factor or argument affecting government investment decisions; although the influence may be present or even decisive, the evidence is too inferential. For this reason, it is accepted that the national nuclear reactor developmental programs received large amounts of government, utility, and (where applicable) private funding and that these played an important role in the economic and political calculus underlying the continuing investment decisions.

The Developmental Periods

Different stages in the developmental process present conditions that suggest different problems or aspects in the development of a technology. A changing set of political, economic, and technical conditions would be expected during the course of the development—that is, from its inception to the final diffusion decision. For current purposes, three distinct periods may be identified: the development period,[59] the demonstration period, and the dissemination period. The first period (primarily described in Chapter 5) is characterized in this analysis by the initiation of the nuclear reactor development program as well as by the early development and testing of nuclear reactors with low thermal or electricity ratings. The second period (Chapter 6) is defined by the demonstration of larger reactors (in the neighborhood of 100 MW[e]). The dissemination period (Chapter

59 The terms "developmental process" or "developmental experience" should not be confused with the "development period," which is a stage in the overall process and describes a bounded period.

7) is when a nation's decisionmakers choose to accept or reject the reactor that had been developed and demonstrated during the previous two periods. Although these definitions are offered in the context of nuclear reactor development, a similar delineation of the developmental process can be made for a more general model.

Given that different conditions prevail during these three periods, one might expect to observe a changing hierarchy of the participating institutions and their objectives as reflected in altered emphases on the different objectives and attributes.[60] For example, during the development period, one could hypothesize that the attributes and institutions contributing to the objective of basic scientific research would be more prevalent than those whose emphases would pertain to obtaining or disseminating an efficient technology. The transition between the periods is particularly important because it suggests how readily the development can move to new objectives or is bound to old ones.

The explicit recognition of distinct periods inserts an element of developmental and institutional dynamics in the analytic framework. It also permits greater discrimination when drawing policy observations. Conditions that might be seen as encouraging the achievement of an objective during one period could also be seen as damaging to that same objective during another, or perhaps a certain objective might be more incongruous in the overall developmental pattern during one period than another.

Relationship of Attributes to Objectives over Time

In this analytic structure, the multiple objectives and the institutions that pursue them as organizational goals are juxtaposed to define the national developmental program. In turn, the objectives are defined by the national attributes. The presence of specific attributes is reflected in the relative importance of particular objectives in the overall national development program.

---

60 Lambright and Teich, "Technology Transfers . . . ," pay special attention to the changing roles of the institutional actors as the developmental process progresses.

For instance, the presence and persuasiveness of an active scientific community at a given time could be manifested by the relative hegemony of the scientific research for science's sake objective at the same time (other things being equal), or a high level of resources invested might invoke the national prestige objective. This functional relationship can conveniently be displayed in a matrix format suggesting the possible interactions between objectives and attributes, as in Figure 2-1. The objective columns are largely couched in terms of organizational goals, and the attribute rows are cast in terms of technology development variables. The matrix graphically depicts the question posed in Chapter 1 addressing what economic, institutional, and political forces shape government-sponsored technology RD&D programs and, conversely, what effect scientific research and technology development programs have upon a nation's RD&D policies.

Attributes of Technology Development / Objectives of Technology Development	1. Active Scientific Community	2. Integrated Tech. Del. System	3. Concomitant Military Program	4. Early Starter	5. National R&D Heritage	6. Multiple Technology	7. Resources Invested
A. Science for Science's Sake							
B. National Prestige							
C. Development of Efficient Technology							
D. Political Equity							

Figure 2-1. Framework for Technological Development: Objectives/Attributes Matrix

Thus, the matrix permits inquiry into the nature of a technology development and which of its aspects have the most significant effects on a given policy objective or set of objectives.

As just stated, over the course of the development one would expect to observe different relative weightings among the attributes and objectives. For this reason, the following case studies are presented in an explicitly comparative manner. Rather than sequentially detailing the six individual national reactor developmental experiences, the work is organized by the three developmental periods.

Part II:
The Comparative Analyses of National Reactor Programs

3

The Initial Decision to Develop Nuclear Research Programs

In considering the geneses and progressions of the national atomic reactor research programs, it is particularly instructive to observe how the different nations surveyed here shared a common set of motivations and constraints during their respective initial research periods. Therefore, rather than beginning the analyses of six national reactor development programs by individual nations, this chapter will compare the national programs as they were affected by similar circumstances and pressures. That is, a shared context or narrative fabric is described here out of which the threads of the individual national experiences will momentarily be drawn.

Initial Motivations and Constraints

In general, the original decision to develop nuclear research programs and later power reactors is due to three stimuli: security needs, the perceived need for power resources, and national prestige. Other objectives may of course emerge later in the development. Both the Soviet and U.S. nuclear reactor programs were founded during "crash" wartime projects, the French and British were publicly motivated by their perception of an imminent shortage in energy resources, and some Third World nations have chosen nuclear reactors as political and technical symbols of their emergence. Either individually or in combination, these three stimuli led the various governments to initiate nuclear power research and shaped their programs in fundamental ways. An additional factor also affected the initial design choices of virtually every country—a factor that is more technical than political or economic—the availability of enriched uranium.

Nuclear Weapons Systems

The Soviet Union, the United States, the United Kingdom, and Canada all devoted their early nuclear research efforts to the development of nuclear explosives as part of their wartime research program during World War II.[1] The last three began within the framework of the wartime Atlantic alliance against Nazi Germany and continued into the exigencies of the postwar period with the notable exception of Canada[2] The Soviets similarly pursued atomic bomb research during the war but not as

---

1 For the wartime development in the Soviet Union, see Arnold Kramish, *Atomic Energy in the Soviet Union* (Palo Alto: Stanford University Press, 1959). The U.S. effort is related by Leslie R. Groves, *Now it Can Be Told: The Story of the Manhattan Project* (New York: Harper and Brothers, 1962), who commanded the Manhattan Project; the definitive account is by Richard G. Hewlett and Oscar E. Anderson, Jr., *A History of the United States Atomic Energy Commission: The New World, 1939-1946* (University Park: Pennsylvania State University Press, 1962), vol. 1 The U.K. research is chronicled by Margaret Gowing, *Britain and Atomic Energy, 1939-1945* (New York: Macmillan, 1964). Wilfred Eggleston, *Canada's Nuclear Story* (Toronto: Clarke, Irwin, and Co., 1965), details the Canadian role in the Allied wartime research effort.

2 Alone among the major wartime allies, Canada has explicitly rejected the development of nuclear weaponry; for possible explanations, see Leonard Beaton and John Maddox, "Canada," in *The Spread of Nuclear Weapons* (London: Chatto and Windus for the Institute for Strategic Studies (1962) ch. 5; or Jon B. McLin, *Canada's Changing Policy, 1957-1963* (Baltimore: The Johns Hopkins University Press, 1967).

part of the allied war effort. In the case of France, which was not a member of the wartime research effort,[3] the initial motivation for a weapon system is less obvious. The postwar French explicitly rejected nuclear weapons in a speech by the French representative to the United Nations in June 1946: "I am authorized to say that the goals the French Government has assigned to the research of its scientists and technicans are purely peaceful."[4] However pacific these statements, even the most defensive of the French nuclear apologists concedes that at least part of the initial French effort to develop nuclear energy capabilities was driven by a desire within the government for nuclear weapons.[5] The influence of this initially covert strategy on the development of the French nuclear program is much less subtle.

The only nation in the sample that developed nuclear reactors without the benefit or assistance of a weapons program was, of course, the Federal Republic of Germany. The reasons are obvious. West Germany unilaterally renounced the manufacture or testing of nuclear weapons as part of its entry into the Western European union in May 1955. An even greater obstacle to West Germany's development of a nuclear capability was the general belief that such a development would be a profoundly destabilizing political event and almost certain to provoke armed Soviet intervention.[6]

The only nation whose power reactor program was influenced by the development of a nuclear propulsion system was the United States. The U.S. Navy emphasized the development of a nuclear propulsion system for both its aircraft carriers and its submarines. Only as an afterthought and as a justification for its

---

3 Many French scientists, of course, lent their skills to the wartime research effort Their experience was later of substantial value to the French nuclear program.

4 Quoted in Scheinman, *Atomic Energy Policy in France . . .*, p. 20.

5 See, for example, Bertrand Goldschmidt, "The French Atomic Energy Program," *Bulletin of the Atomic Scientists* 18, 7 (September 1962): pp. 39-42. Goldschmidt was a leading French nuclear scientist who had worked on the Canadian program during the war before returning to France to work on the French program.

6 See Henry A. Kissinger, *The Troubled Partnership* (New York: McGraw-Hill for the Council on Foreign Relations, 1965), pp. 142-143; for an earlier but more detailed discussion, see Beaton and Maddox, "Germany," *The Spread of Nuclear Weapons*, ch. 6.

program did the Navy add a power production capability to its specifications.[7] The Soviet Union developed a nuclear propulsion system for its icebreaker, the *Lenin*, but development did not begin on the *Lenin* until 1956, well after the groundwork for the Soviet reactor system had been established.[8] It did not have the important effect on the Soviet reactor program that the U.S. Navy's research agenda had on the American power reactor program.

The influence on power reactor research of the early impetus toward nuclear weapons or propulsion systems is not difficult to identify. In the United States, the Office of Naval Research first expressed a research interest in the possibility of using atomic energy for submarine propulsion in 1939.[9] In 1948, Westinghouse became the prime contractor for the development and testing of the Mark I nuclear reactor, a full-scale, land-based prototype of what was to become the Nautilus family of pressurized water reactors (PWR) for submarine propulsion.[10] Government physicists at the Argonne National Laboratory also contributed to this objective. By 1950, this program was considered a major part of the U.S. reactor development program.[11] The Mark I reactor went into full-power operation in June 1953, "the world's first fully-engineered nuclear reactor capable of producing practical amounts of energy on a sustained and reliable basis."[12] The research efforts that had gone into the Mark I program gave the light water reactor (LWR) a clear advantage over competing reactor alternatives that were being concurrently developed in the United States, such as the experimental breeder reactor, the homogeneous reactor, or the sodium-cooled reactor. In particular, it gave Westinghouse and the PWR a practical advantage over other commercial producers and LWRs such as General Electric

---

[7] See Hewlett and Duncan, *Nuclear Navy*.

[8] A.M. Petrosyants, *From Scientific Search to Atomic Industry* (Danville, Ill.: Interstate, 1975). Petrosyants is the Chairman, USSR State Committee on Use of Atomic Energy.

[9] Greenberg, *The Politics of Pure Science*, pp. 71-72.

[10] Richard G. Hewlett and Francis Duncan, *A History of the U.S. Atomic Energy Commission: Atomic Shield, 1947-1952* (University Park: Pennsylvania State University Press, 1969), vol. 2, p. 207.

[11] Ibid., pp. 217-219.

[12] Hewlett and Duncan, *Nuclear Navy*, p. 186.

and the boiling water reactor (BWR). As Weinberg commented at the time, "The choice of water cooling and moderation for PWR was dictated by the requirement that the reactor demonstrate reliable nuclear power rather than cheap nuclear power."[13]

Although each of the various prototype reactors had its advantages and proponents, the advanced work done under the Navy's propulsion program on the PWR meant that it had fewer technical uncertainties than any other reactor type. It was no surprise in 1954 when the contract for the first commercial reactor built in the United States, the Shippingport reactor in Pittsburgh, Pennsylvania, was awarded to Westinghouse; the Shippingport design was basically an enlargement of Westinghouse's *Nautilus* propropulsion unit.

The desire for nuclear weaponry had an important effect on the British and French nuclear reactor programs. In England, the decision to build an atomic bomb was made soon after the end of the war. Therefore, the initial British reactor decision was predicated upon the need to produce weapon grade fissile materials, which meant constructing either a gaseous-diffusion plant for producing enriched, weapons grade U-235 or a graphite pile for plutonium (Pu). In September 1945, Sir James Chadwick (Chairman of the subcommittee on atomic energy of the Government's Advisory Council on Scientific Policy) wrote that "the military applications of atomic energy obviously made a production plant 'of our own' essential for the defense of the United Kingdom and the Commonwealth. 'I believe there can be only one opinion on this question'."[14] Foreign Minister Ernest Bevin is reported to have phrased that opinion more colloquially: "Let's forget about the Baroosh [the American-proposed Baruch plan] and get on with making the fissile [sic]."[15] Once the bomb decision had been made, the major debate was over whether the bomb should use U-235 or plutonium. Chadwick was terse: "The decision has been taken to make bombs and therefore we must make

[13] Alvin Weinberg, "Power Reactors," *Scientific American* 191, 6 (December 1954):31.

[14] Margaret Gowing, *Independence and Deterrence: Britain and Atomic Energy, 1945-1952* (New York: St. Martin's Press, 1967), vol. 1, p. 164.

[15] Ibid., p. 164.

plutonium."[16] Indeed, the weapon's choice dictated plutonium. The critical mass of Pu needed for an atomic bomb was thought to be much less than that of U-235, and its explosive efficiency was greater; British scientists "calculated that a bomb would need ten times as much U-235 as plutonium to produce half the TNT equivalent."[17] The Pu plant would also cost much less than a U-235 enrichment facility. These considerations led the British government in 1947 to choose a Pu-producing, graphite moderated, air cooled reactor to be located at Windscale. The decision for the plutonium-producing facility—as opposed to one that would produce enriched uranium—all but guaranteed that the British would be forced to adapt its power reactor prototype from a family of reactors that did not require enriched U-235.

The first question facing French nuclear authorities after the war was whether France should attempt to focus its nuclear materials and personnel on industrial uses or undertake the production of fissionable material. Rather than be dependent on foreign sources for the supply of fissile materials, France—like Britain—chose the latter alternative. The interesting policy question was the French choice between a U-235 or a Pu-producing facility. Publicly, on the basis of "technological and cost factors, and in the belief that plutonium was the key to an atomic future, the CEA [the Commissariat à l'Energie Atomique, the French Atomic Energy Commission] opted for the production of plutonium"[18] and built a small, graphite moderated, air cooled (at atmospheric pressure) facility at Marcoule designed to produce 12 kilograms of plutonium annually.[19] As Scheinman points out, the "material and technical situation in France in 1946 substantially narrowed the range of the policy alternatives available in atomic development."[20] However, as Scheinman continues, "at the time, the only proven use of plutonium was for military purposes. This would seem to support the contention that the Five-

---

[16] Ibid., p. 176.

[17] Ibid., p. 165.

[18] Scheinman, *Atomic Energy Policy in France . . .*, p. 67.

[19] Francois Perrin, "Atomic Energy in France," *Bulletin of the Atomic Scientists* 11, 3 (March 1955):93. Perrin was the Administrator of the CEA when this paper was published.

[20] Quoted in Scheinman, *Atomic Energy Policy in France . . .*, p. 20.

Year Plan had other than peaceful ends in view."[21] Bertrand Goldschmidt, who was the leading French authority on the production of plutonium, later noted that "the 1952 five-year plan made no mention of the possible use of plutonium in military applications, but it is certain that this eventuality was present—if not predominant—in the minds of those who suggested the five-year plan and encouraged its adoption."[22] As with the British, this choice for a plutonium facility fundamentally affected the design of the initial French power reactors by limiting the French options to reactors not requiring enriched uranium.

The Soviet example is less certain because of the lack of information. However, it is probably safe to speculate that the Russian crash program to develop nuclear weapons in the late 1940s almost certainly preempted reactor research resources,[23] especially after it was clear that the United States would not share its nuclear secrets with any of its wartime allies. This refusal all but guaranteed that the Soviets would develop their reactor program as a poor stepchild to their nuclear weapons.[24] However, even given the Soviet emphasis on nuclear explosives, the first of which was exploded in September 1949, considerable resources must have been expended on nuclear reactor research because that same year a decision was made to build a small power demonstration reactor at Obninsk.[25] The Obninsk reactor, which went into operation on 27 June 1954, was the first operational electricity generating nuclear power station in the world.[26] Unfortunately, we have no direct access to the decisionmaking processes that led to the decision or the construction of this facility. For example, it is impossible to discuss what reactor options the Soviets might have considered; therefore, it is difficult to speculate how the early Soviet emphasis on the development of

---

[21] Ibid., p. 88.

[22] Goldschmidt, "The French Atomic Energy Program," p. 40.

[23] See A. Petrosyants, "Nuclear Energy—A Productive Force," *New Times* 11 (1971):18.

[24] Of course, U.S. reluctance to share its nuclear secrets can also be seen as a major motivation behind the French and British programs to develop indigenous nuclear weapons and reactor programs.

[25] See Petrosyants, *From Scientific Search to Atomic Industry*, pp. 83-88.

[26] See Kramish, *Atomic Energy in the Soviet Union*, p. 144.

nuclear bombs might have affected the development of the nuclear power reactor.

The terms of West German sovereignty in 1955 forbade any nuclear weapons research. Furthermore, the 1954 Eisenhower Atoms for Peace Plan and the German membership in EURATOM provided a secure source of fissile materials for the nascent German atomic research community. Thus, the initial nuclear questions that confronted Britain and France were never critical obstacles to the German nuclear planners. "Production of high-grade plutonium was never an issue. National independence was not considersd a necessity."[27]

The final example of the influence of military programs on reactor development is the Canadian experience. As part of the cooperative allied wartime effort to manufacture an atomic bomb, the Canadians were assigned the responsibility of pursuing "heavy water" or deuterium oxide (D_20) research. The entire European supply of D_20 was transferred from England to Canada along with a number of scientists, and in September 1942, "preparations were begun for the establishment of Canada's first large-scale nuclear laboratory."[28] As a result of the wartime research program, the postwar Canadian nuclear program benefited from its exposure to heavy water as both a moderator and a coolant. At the same time, the United States chose to keep its nuclear secrets and technology from the Canadian research program, thus depriving Canada of any working knowledge of the uranium enrichment process. These two features of their wartime research experience were to have profound effects on Canada's nuclear reactor program.

Power Resources

The second major influence on a nation's initial nuclear reactor decision is the perceived need for new energy

---

[27] Dr. W. Finke, "Survey of F.R.G.: Research Activities Directed Towards the Nuclear Generating Programme," *Nuclear Engineering* 13, 148 (September 1968):743.

[28] Eggleston, *Canada's Nuclear Story*, p. 53.

resources—specifically electric power and more important just how pressing or immediate that need appeared to be. The case is perhaps clearest in France. Perrin, the High Commissioner of the CEA, wrote in 1955, "In France we do not have enough coal, very little oil, and limited hydroelectric possibilities. . . . If we are not then able to build atomic power stations, we will have to use more and more expensive imported coal or oil for power production."[29] Energy statistics cited by Scheinman support Perrin's description: "Of all the members of the Organization for European Economic Cooperation, France has been the biggest net importer of energy."[30] This was even more apparent after the devastation of World War II. The Ruhr coal fields would not be French, and the French had ambitious plans for postwar industrialization and modernization. Almost certainly, then, the fear of diminishing energy resources in the face of increased demand was a contributing factor in the French decision to develop nuclear facilities.

This motivation can perhaps best be observed by the participation of French nuclear scientists, many of whom were Socialists and even Communist party members during the late 1940s, who would have refused to participate in a program whose primary orientation was the development of nuclear weapons. Only after Frédéric Joliot-Curie, an admitted Communist party member, was removed from his position as High Commissioner of the CEA in 1950 did the French program manifest its movement toward producing nuclear weapons.[31] In any case, just how important the perceived need for additional energy sources was a preliminary motivating factor underlying the French decision to develop its nuclear resources is difficult to ascertain, but it is clear that the military mission had supplanted it by the mid-1950s.[32] As one might expect, this exchange of priorities had an effect on the

[29] Perrin, "Atomic Energy in France," p. 93.

[30] Scheinman, *Atomic Energy Policy in France . . .*, p. 60.

[31] The removal of Joliot-Curie was an event of great importance in France; see Scheinman, "The Formative Years: 1946-1961" in *Atomic Energy Policy in France . . .*, ch. 2. Also see Beaton and Maddox, "France," *The Spread of Nuclear Weapons*, ch. 4.

[32] The primary need for nuclear power as an energy source was still cited in 1962; see Bertrand Goldschmidt, "The French Atomic Energy Program," *Bulletin of the Atomic Scientists* 18, 8 (October 1962):46-48.

development of the nuclear power reactor, almost certainly retarding its development.

The British case is less clear cut, for U.K. energy resources—mainly coal—were more certain. In England, the alternative energy or power resources debate revolved around the economic advantages or disadvantages of atomic power versus coal. The prevalent attitude was expressed by Gowing: "In general, unless there was the prospect that atomic energy might ultimately be capable of satisfying an appreciable part of the country's heat and power requirements, it would not be worth developing on an industrial scale."[33] Its primary value at the time seemed to be as a replacement for other energy sources besides coal, for example, oil. However, as the nuclear weapons arguments began to lose their persuasiveness in the early 1950s, the Conservative government began to emphasize the value of atomic energy as an additional and economic source of power. Duncan Sandys, then Minister of Supply, offered the following guarded assessment to the House of Commons (26 January 1953):

> It is too early to say with any precision how soon electricity generated from atomic energy will be available on a significant scale for industrial purposes. It should not be imagined certainly that nuclear reactors will in the near future supersede existing methods of producing electric power. But if . . . the technical problems involved are successfully solved and these new methods prove to be economical, there is no reason why nuclear reactors should not before very long provide a useful additional source of industrial power.[34]

Sandys was not far amiss in his prediction. In February 1953, the United Kingdom's Atomic Energy Authority authorized a dual purpose (plutonium and power production) reactor to be built at Calder Hall. In May 1956, the reactor went critical and by July was delivering electrical power to its neighboring plant, Windscale. In October, Queen Elizabeth II opened the station officially and, "for the first time anywhere in the world, nuclear power had been used to generate the electricity used in factories and

---

[33] Gowing, *Independence and Deterrence*, p. 238.

[34] See "British Debates for Atomic Power," *Bulletin of the Atomic Scientists* 9, 4 (May 1953):141.

homes."[35] This was over a year before the first U.S. nuclear plant at Shippingport began to generate power.[36]

Sir Christopher Hinton, then Chairman of Britain's Central Electricity Generating Board, wrote that it "would seem reasonable to say that, starting from nothing in 1946, the British are, temporarily at any rate, ahead of the U.S. in the industrial-reactor field."[37] The reason for this rapid advancement is a complex amalgam of research strategies and priorities, but a significant part must be attributed to England's greater and more immediate perceived need for nuclear power to supplement its coal and oil resources. Hinton commented:[38]

> The real reason for the present relative positions of the U.S. and the United Kingdom in the industrial application of nuclear power seems to lie in the well-known relation between necessity and invention. British supplies of coal are running short. All of our oil is imported. From the outset we realized that we would need nuclear power and need it soon. . . . There is no reason to believe that Americans could not have organized with equal success, but you did not have the same need.

Concerns for additional energy resources were not strong motivating factors in the United States, Canada, West Germany, and apparently the Soviet Union. During the early stages of reactor research, the United States was well endowed with supplies of domestic coal and petroleum for its energy needs. Canada was well supplied with hydroelectric facilities, and the Federal Republic of Germany was a net exporter of energy supplies. In 1956, when West Germany first began to consider research on nuclear power reactors, the Minister for Atomic Questions, Franz Josef Strauss, "made it clear that Germany had plenty of time to catch up in nuclear research before nuclear power became a necessary and important source of domestic energy production."[39] In addition, the initial German decision was made

---

35 Sir Christopher Hinton, "Atomic Power in Britain," *Scientific American* 198, 3 (March 1958):29.

36 Stephen E. Schattmann, "Britain's Nuclear Power Program," *Science* 127, 3313 (June 27, 1958):1489.

37 Hinton, "Atomic Power in Britain," p. 35.

38 Ibid., p. 35.

39 Henry R. Nau, *National Politics and International Technology: Nuclear Reac-*

in the mid-1950s, much later than the analogous French and British decisions, when "there was a world-wide affluence of coal and therefore it was no longer convincing to argue that a primary energy gap was imminent."[40] Information on the Soviet Union's need for alternative energy sources is sparse, but there is indirect evidence that the Russian nuclear research program on reactors was later driven by this requirement, especially for the western portions of the country.

The influence of the projected lack of energy resources is important. The relative abundance of power sources implies that a given reactor type need not be developed on a high priority basis. This freedom permitted the United States and the Soviet Union to develop a number of reactor types concurrently, providing the superpowers with a varied menu from which to choose the preferred reactor type.

The Germans enjoyed a similar luxury, although for different reasons. Starting nuclear reactor development as late as the Germans did was possible because Germany had sufficient power to meet its needs. This allowed West Germany to wait and then choose among alternative reactor types that had already been largely developed. The Canadians did not use their relative advantage in power resources to develop a variety of reactor types. Rather, they chose to develop the Canadian reactor model in a studied, careful, and lengthy progression. The perceived need of the French and British to develop supplemental sources of electric power impelled both to make decisions on reactor types as soon as possible. In retrospect, this early forced decision had deleterious effects on both nations' nuclear industries. France and Britain were "locked" into nuclear reactor technologies that were to become quickly outmoded in light of advancing reactor types and economies.

---

tor Development in Western Europe (Baltimore: The Johns Hopkins University Press, 1974), p. 85.

40 Robert Gerwin, *Atoms in Germany* (Dusseldorf: Econ-Verlag for the Bundesminister für Wiessenschaftliche Forschung, 1964), p. 28.

National Prestige

The final reason for initiating a nuclear research program was national prestige. France was the most obvious example. Defeated and occupied in World War II and treated as a second-rate member of the U.S.-U.K. alliance, France was caught "between aspirations and capabilities. Only two real alternatives existed: either France had to relinquish her pretensions, reconsider her goals and reconcile herself to new circumstances, or she had to strive to reassert herself materially in the rank of Great Power nations."[41] Nuclear energy was seen by the French government as both a symbol of great nation status and a basis for French industrialization.[42] Given the French scientific heritage[43] and these conditions, it was a certainty that France would develop a nuclear power reactor system and probably nuclear weapons as a matter of national prestige and pride. Much the same motivation can be attributed to West Germany close to a decade later. Having largely recovered from the destruction—but not the stigma—of war, Germany turned to nuclear energy as a prestigious manifestation of its industrial strength. Atomic Minister Strauss specifically addressed himself to this issue at the inaugural meeting of the German Atomic Commission in 1956:

> For it is not a matter of military or political power. Nor is it a matter of prestige. But it is indeed a matter of asserting and securing the place of the German people among the industrial nations, a place which we have regained again with so much effort.[44]

The British clearly viewed their reactor developments as part of national pride and international technological preeminence. Queen Elizabeth II spoke of Calder Hall as "something new we have to offer to the peoples of the underdeveloped and less fortunate areas of the world, who will continue to look to us for

---

41 Scheinman, *Atomic Energy Policy in France . . .*, p. 216.

42 See Robert Gilpin, *France in the Age of the Scientific State* (Princeton: Princeton University Press, 1968), pp. 282-288.

43 French scientists had applied in early 1939 for the first patents covering the general principles of nuclear reactors, including moderating materials.

44 Quoted in Nau, *National Politics and International Technology*, p. 72.

assistance and example as they have done in the past,"[45] while Hinton heralded Calder's opening as that "day the United Kingdom atomic energy program, which had been through a good deal of depressing weather, broke through into the sunshine of international prominence."[46]

In the Soviet Union, nuclear weapons and later nuclear power production were more than a matter of prestige; they were actively engaged in and advertised as demonstrations of the superiority of the Communist ideology and the greatness of the Soviet system. The Soviet government was quick to highlight its early advances in nuclear technology and sponsored carefully escorted tours by Western observers of their facilities,[47] an opportunity particularly available to it because of the relatively closed society. Although nowhere near as blatantly nor consciously, the United States had developed nuclear energy as a well-advertised symbol of its technologically advanced industrial society. The motivation is less clear for Canada, even though a certain element of national pride probably went into the Canadian decision to develop an indigenous nuclear power reactor system while consciously foregoing the nuclear weapons system.[48]

In the United States and Canada, national pride seems to have played an important role in determining the decision to develop nuclear energy, but not as critical a role as in France, Britain, the Soviet Union, or West Germany. Other considerations were obviously present—economics and the continuation of wartime-initiated research to name two—but national prestige cannot be excluded.

The actual influence of prestige on national development programs is difficult to isolate. The West German government provided previously unheard of amounts of funding into R&D, an area that was traditionally dominated by private industry and, secondarily, by the individual states. Government support was

---

[45] Quoted in *Nuclear Engineering* 1, 8 (December 1955):822.

[46] Hinton, "Atomic Power in Britain," p. 29.

[47] For example of an early tour, see "W.H. Zinn Evaluates Russian Reactor Progress," *Nucleonics* 17, 3 (March 1959):104-107; a later but similar report is provided by Robert Gillette, "Nuclear Power in the USSR: American Visitors Find Surprises," *Science* 173, 5001 (September 10, 1971):1003-1006.

[48] See Beaton and Maddox, "Canada," p. 107.

particularly remarkable given that there was no apparent perception of an imminent energy shortage. For this reason, one can posit that matters of national prestige were important factors influencing the central government's enthusiastic support and sponsorship of nuclear research.

The influence and effect of national prestige can most clearly be seen in the French, British, and Soviet examples. For the French and the British, nuclear reactors were seen as a way to reaffirm their great power status, which was then tarnished. In both nations, the element of national prestige underlying their reactor development programs was most clearly manifested by their insistence upon domestic technology and fissile materials. Both France and Britain made early policy decisions not to accept or solicit foreign nuclear technology that might restrict their freedom of development or use. Even after the restrictions of the McMahon Act were lessened with Eisenhower's Atoms for Peace (1954) proposals and the EURATOM Agreements (1956), pressures of national prestige restrained both nations from importing foreign technologies until later in the 1960s when their domestic nuclear technology was indisputably lagging and user complaints were no longer satisfied by appeals to patriotism. In the Soviet Union, reactor development was emphasized as a demonstration of the technological prowess of the Soviet economy. It is hardly coincidental that the Obninsk reactor achieved criticality in time for the convening of the 1955 Geneva Conference on the Peaceful Uses of Atomic Energy.

Enriched Uranium

A final important influence on the initial design decisions for nuclear power reactors is more technical and less political. This was the availability of uranium U-235 for enriching reactor fuel. The uranium enrichment process that the Manhattan Project perfected during the wartime research program was a closely guarded American secret because of its implications for the proliferation of nuclear weapons. Even today, "the only important vestige of the proprietary and operational features of the Atomic Energy Act of 1946 is the United States government monopoly of the technology and facilities (gaseous diffusion plants) for producing enriched uranium."[49] Indeed, the U.S. government later prevailed upon the governments of the Netherlands and West Germany to classify their national and cooperative research on the gas centrifuge uranium enrichment process and has forbidden U.S. private research in this area except under government contract.

American reluctance or outright refusal to provide enriched uranium had a fundamental effect on the British, French, and Canadian nuclear reactor programs and, in fact, on their entire nuclear research programs. As discussed above, both the British and French were confronted with the necessity to make an early decision between building a less expensive plutonium production plant or a technically very difficult U-235 separation plant, or relying on the United States as a source of enriched uranium. Their choice to construct facilities for domestically producing plutonium dictated that their first generation reactors would have to be fueled with "natural" or unenriched uranium. This, in turn, meant that the light water cooled and moderated family of reactors, which utilized uranium-238 enriched with U-235, was not an available option to them. Their reactor designs would have to be more "efficient;" the fuel cladding, coolant tubes, and the water cooling annuluses must be thin and have low neutron absorption. But this necessarily imposed size, capacity, and cost restrictions that would later be debilitating. Both the French and British developed a graphite moderated, air or carbon dioxide cooled nuclear reactor as their first generation model. It was not until the enrichment facilities at Capenhurst (1962) and Pierelatte (1967) were functioning that England and France could supply their own enriched uranium and develop other reactor types. By that time, both nations had developed reactors based on unenriched uranium fuels and the technological "sunk costs" made the transition to enriched uranium reactors formidable.

Canada similarly found itself without access to enriched uranium and was thereby limited in the number of reactor types

---

49 Lawrence Scheinman, "Security and a Transnational System: The Case of Nuclear Energy," in Robert O. Keohane and Joseph S. Nye, Jr., eds., *Transnational Relations and World Politics* (Cambridge, Mass.: Harvard University Press, 1972), p. 278.

it could develop. Canada's abundance of natural uranium (U-238) ore and her wartime research emphasis on heavy water led the Canadian nuclear reactor technicians to develop a power reactor fueled with unenriched uranium and based upon heavy water as both a coolant and a moderator.

The early Soviet reactor, which was to serve as prototype for the later power reactors, was a graphite moderated, water cooled system. This might suggest the Soviet Union was similarly constrained by a lack of enriched uranium even though, of course, the Soviets had developed their own enrichment process as part of their weapons research program. In fact, however, the Obninsk power reactor did operate on which Soviet nuclear technicians referred to as "poorly enriched" uranium fuel[50] because they thought that

> It was impossible to obtain a sufficiently high thermodynamic efficiency [utilizing natural uranium]. At the same time, utilization of enriched uranium succeeded in increasing the reactor's specific power, in maintaining comparatively small design dimensions and in increasing the strength characteristics of the materials used. These were the decisive factors in choosing the nuclear fuel for the reactor of the first nuclear power plant.[51]

It appears that the Soviets chose the graphite moderated plant more for its simplicity of construction and operation than because of a lack of enriched uranium.

The West Germans began their nuclear reactor program a full decade after the British and French. By this time, the United States was prepared to guarantee European nations enriched uranium under the auspices of EURATOM. When the Germans decided in 1960 to build a demonstration plant, the bidding was narrowed down to a British gas-graphite plant rated at 340 MW(e) and a General Electric BWR reactor rated at 250 MW(E). The contract was awarded to GE. Nau comments that "the existence of this [EURATOM] program altered expectations in favor of American reactors from 1958 on. Europeans were assured of a hand in research results in the United States, and U.S.

---

50 Petrosyants, *From Scientific Search to Atomic Industry*, p. 87.

51 Helen Knorre, "Frontiers of Technology: U.S.S.R.," *Nuclear Engineering* 9, 100 (September 1964):317.

guarantees of enriched uranium supplies removed a primary obstacle to light water developments in Europe."[52]

Summary

Four major factors—national security, perceived need for additional energy resources, national prestige, and the availability of enriched uranium—had major effects on the initial motivations for and designs of nuclear reactors and the attendant development and strategies of six industrialized nations. The U.S. program was a continuation of its wartime research projects and the submarine propulsion program. These developmental efforts expended on the Mark I and its resulting reliability would strongly bias the U.S. nuclear reactor RD&D programs toward the PWR version of the LWR as the leading U.S. reactor when the construction of the first commercial power reactor was imminent. The Soviet Union also benefited from a continuation of its wartime research experience. Although precise data are lacking, the Soviet program apparently combined its access to enriched uranium with the comparative simplicity and dependability of a graphite reactor for the first Soviet power reactor prototype.

The French and British programs were similar to each other, even though the English had the advantage of cooperative wartime research with the Americans and Canadians. Both nations developed national reactor programs as an outgrowth of atomic national security programs (a feature shared with the United States and the Soviets) and for national prestige (much more obvious than similar attitudes manifested in the United States or even the Soviet Union). Neither country had access to enriched uranium. In addition, Great Britain and France were motivated by a perceived need to augment their national power sources. This convergence of factors determined that each nation's nuclear research program would initially be dedicated to producing plutonium of weapon grade for nuclear weapons, that they would probably have little choice among competing reactor designs except to develop graphite moderated reactors, and that

---

[52] Nau, *National Politics and International Technology*, p. 145.

each nation would make a concerted (perhaps too concerted) effort to develop a nuclear power system.

Canada developed its reactor program as a continuation of its wartime research on the physical properties of heavy water; the program was shaped by the exclusion from the American uranium enrichment processes or materials, the presence of alternative power sources, the lack of a weapons program and its attendant requirements (i.e., plutonium), and the abundance of natural uranium in Canada. This particular combination of events led the Canadians to develop the heavy water cooled and moderated CANDU reactor. West Germany was primarily motivated to obtain nuclear power reactors because of national prestige and industrial expansion; power requirements were secondary and military applications nonexistent. Because of its late participation, West Germany was able to avail itself of largely demonstrated nuclear technologies; furthermore, American enriched uranium was available through the EURATOM joint development program. Therefore, Germany was able to choose among competing reactor types with much geater certainty than its European neighbors. By 1960, many of the uncertainties among and within the various technologies and designs had been identified and eliminated in many cases.

4

Early Organizational Actors

Chapter 3 delineated a common set of political and technological motivations and constraints that mutually and fundamentally shaped the initial decisions of the six nations to begin nuclear research, and special attention was paid to the issue of nuclear reactors. This chapter discusses the commonalities observed in their institutional arrangements. Important parallels may be noted in how these nations organized their nuclear research.

Technology and Organizations

There is a school of thought that claims technological phenomena and their economics cannot be understood without an appreciation of the politics and society that determine them. For example, Gilpin writes, "Politics determine the framework of economic activity and channel it in directions which tend to serve the political objectives of dominant political groups and organizations.

Throughout history each successive hegemonic power has organized economic space in terms of its own interests and purposes."[1] McDermott implies the primacy of politics when he writes, "Technology should be considered as an institutional system" that reveals "systematic and decisive social changes, especially the political and cultural tendencies, that follow the widespread application of advanced technological systems."[2] Nau's thesis is that domestic politics, not economics or technology, dominate the international exchange of nuclear technologies.[3] It is not central to this research to determine which force—political, economic, or technological—was primarily responsible for shaping the development and dissemination of nuclear reactor research and programs. Each of them and the interactions among them were both necessary and complex.[4] The technological development of nuclear power reactors was not an isolated phenomenon driven solely by scientific and technological advances; domestic and international politics played a substantial role in the RD&D processes, and a variety of institutions and actors emerged as important players in different nations.

This chapter discusses a number of functional topics and how they were comparable or differed from nation to nation.[5] In this way the national, cultural, and societal similarities and distinctions are pointed out more readily than if the nations were individually described. The influence of the respective systems and institutions upon the various national programs is posited and examined at the end of the chapter.

The common topics or areas of organizational analysis are *the role of the central government*, especially the executive branch,

---

[1] Robert Gilpin, "The Politics of Transnational Economic Relations," in Keohane and Nye, eds., *Transnational Relations and World Politics*, p. 53.

[2] John McDermott, "Technology—The Opiate of the Intellectuals," in Albert H. Teich, ed., *Technology and Man's Future* (New York: St. Martin's Press, 1972), p. 171.

[3] Nau, *National Politics and International Technology*, passim.

[4] For example, see Emmanuel G. Mesthene, *Technology Change* (New York: New American Library, 1970).

[5] The primary government agencies dealing with nuclear research and regulation in the United States, Great Britain, France, Canada, Italy, Japan, and EURATOM are described by J.E. Hodgetts, *Administering the Atom for Peace* (New York: Atherton Press, 1964).

which manifested itself in a perceived necessity for government control; *the initial roles of private industry and electric utilities*; and *the roles of other groups*, such as labor unions, universities, scientist lobbies, or environmental groups. These demarcations do not imply that such clear distinctions can be observed in all cases; for example, the Soviet system of government obscures most of these categories. However, these distinctions can generally be made at least on a functional basis. Qualifications are observed where necessary. This discussion of institutional factors details only those organizations and groups involved in the initial stages of national nuclear research—organizations that had an effect on the early design and development decisions. Later decisions and programs, such as those dealing with the commercialization of nuclear power reactors and their regulation, are treated at greater length in subsequent chapters as they become pertinent. However, it is important to recognize their early contributions to appreciate their later significance.

The Necessity for Government Control

Scheinman has posited four decisive reasons for the early governmental involvement in and control of nuclear research: the wartime weapons research experience; the perceived importance of nuclear energy to national industrial growth; the national and international prestige and status associated with nuclear technology; and the immense cost of nuclear research and development.[6] The first three factors mirror the earlier motivations of the sample of nations who initiated nuclear research. The fourth reason, finances, was an equally critical reason for governmental intervention.

The first and most important rationale for government control grew out of the wartime experience, with the important exception of West Germany. Scheinman has commented, "During World War II and immediately thereafter the prevalence of military considerations, awareness of the dangers of information 'leaks', and only a rudimentary understanding of how atomic energy

[6] Scheinman, "Security and a Transnational System," p. 280.

could safely and effectively be harnessed for peaceful use combined to lead the first nuclear powers to assert total government control over nuclear energy."[7] The United States, Great Britain, and Canada all began their nuclear research under the provision of the 1943 Quebec Conference agreements when concerns of secrecy were paramount. Indeed, security concerns were to hamper and later restrict to the point of termination the flow of nuclear information between the Western Allies.[8] The wartime research efforts had been entirely controlled by the central government so it was only natural that the postwar nuclear research programs would continue to be controlled by the governments, especially as long as weapons production was a primary concern. Although Canada did not pursue nuclear weapons development, the government did choose to remain the central actor in the Canadian reactor development program. Perhaps significantly, the Canadian government was among the first of the nuclear powers to decentralize its nuclear R&D.

The Soviet Union's reactor research program similarly was an outgrowth of its nuclear weapons program. Given the structure of the Soviet political and economic system, there was never any doubt that the central government would be the primary actor in the development of nuclear power. As one might expect, the central government's control of nuclear reactor development and proliferation remained firmer throughout this early period of nuclear research than that of any other nation. However, this distinction was more a matter of degree than of kind.

France did not engage in wartime atomic research even though some French citizens did. Still, it may reasonably be argued that France can be considered as a nation whose early nuclear programs were largely—albeit covertly—driven by a national defense rationale. French scientists, politicians, and administrators had visions of a French nuclear force before 1960; deGaulle's *force de frappe* was only a continuation and overt manifestation of policies initiated and carried out during the

---

[7] Ibid., pp. 276-277.

[8] For different national perspectives of this problem, see Groves, *Now It Can Be Told*; Hewlett and Anderson, *The New World*; Gowing, *Britain and Atomic Energy*; and Eggleston, *Canada's Nuclear Story*.

Fourth Republic.[9] Therefore, it can plausibly be argued that France's efforts to obtain nuclear weapons led to a governmental control that in many respects resembled that of the nations with nuclear research programs during the war.

The one important exception to these examples is the Federal Republic of Germany. Although Nazi Germany had a nuclear research program during World War II, German atomic scientists were not able to marshall the necessary resources and it proved to be inconsequential.[10] The defeat and dismantling of the German industry after the war completely severed any thread of connection between this work and the postwar nuclear programs in West Germany. In 1955, West Germany renounced nuclear weapons, declaring "that the Federal Republic undertakes not to manufacture in its territory any atomic weapons, chemical or biological weapons."[11] The lack of a wartime research experience and the renunciation of developing nuclear weaponry meant that the West German government could not share the same military continuity and motivation that led the other national governments to control their nuclear research. Even though there were other reasons, such as the decentralized nature of West Germany's political system and the fragmented electric utility system, the absence of the government's heavy involvement with military nuclear research is pivotal in understanding the reluctance with which the central government moved to control Germany's nuclear R&D.

The perceived importance of atomic energy as a substitute for diminishing natural resources was a reason often cited for government-sponsored programs dealing with the development of power reactors. Jensen has commented that after the war, much of Europe was particularly concerned about the shortage of power sources:

> The end of the second world war left Europe with a burning need for energy. Energy to give warmth to the thousands who had been made homeless, to repair shattered transport systems, to rebuild national in-

---

[9] This is the thrust of Scheinman's book, *Atomic Energy Policy in France. . . .*

[10] See Gerwin, *Atoms in Germany*, ch. 2, "German Activities During the War," for a brief account.

[11] Quoted in Beaton and Maddox, *The Spread of Nuclear Weapons*, p. 109.

> dustrial structures and to provide a sound basis for economic revival. Energy, in short, was the raw material upon which Europe's recovery depended.[12]

Coal production was about a third of the prewar levels, petroleum imports were down, and all refineries had been destroyed during the war. Because of these circumstances, Britain and France turned to examining nuclear energy as a new potential source of power. The promise of nuclear power caused these governments to invest heavily in creating a viable nuclear power reactor, given that one role of government is assuring (or at least promoting) the economic well-being of its citizens. The premise is particularly applicable where the government dominates the economic market forces, such as the Soviet Union, or manages significant portions of the nation's economy and industries, such as France and Great Britain.

The development of nuclear energy as an alternative source of power was less pressing in the power wealthy United States, Soviet Union, Canada, and West Germany.

The evidence for government intervention and control because of the national prestige accorded nuclear nations is subtle. Still, it is safe to assume that nations perceived nuclear energy as a requisite for national prominence in the international community of nations. Although it is difficult to prove, national governments undoubtedly undertook and controlled nuclear research programs so as to elevate or assure national prominence. Naturally, such a position would have offered beneficial effects to its citizens that the government could not be expected to ignore, thereby providing a rationale for governmental control.[13]

Finally, the large financial cost and technical uncertainties of nuclear research and development, even foregoing the weapons or propulsion systems options, was almost certainly greater than any nongovernmental organization or consortium of organizations could fund. Government intervention was therefore essen-

---

[12] W.G. Jensen, *Nuclear Power* (Cambridge, Eng.: G.T. Foulis, 1969), p. 4.

[13] National decisionmakers have been known to assume and maintain financially and technologically tenuous projects in pursuit of national prestige. For two examples, see Costello and Hughes, *The Concorde Conspiracy*; and Logsdon, *The Decision to Go to the Moon.*

tial if there was to be any nuclear reactor research. In the initial stages of nuclear research, most of the research monies were devoted to weapons systems, clearly a governmental function. However, with only rare exceptions, all of the initial (and a great deal of subsequent) reactor research was also conducted by government laboratories or under government contract because private industries and universities did not have the necessary facilities, manpower, or capital. Also, in the initial stages of nuclear research, private industry and utilities were justifiably hesitant about committing large amounts of capital to the future nuclear industry because of the large financial and technical uncertainties attached to peacetime nuclear energy. Given the technological and financial uncertainties, national decision-makers were correct in assuming the control and direction of nuclear power reactor R&D.

Each of these four influences played an important role in leading the governments to assume control and regulate participation in the initial stages of nuclear reactor R&D to ensure the orderly, expensive development of an efficient technology. Other factors, such as a concern for the environment, were present but clearly played secondary roles until later in the nuclear era.[14] In summary, then, the wartime experience and military implications, the perceived need for supplementary national energy resources, national prestige, and the exorbitant costs attached to nuclear reactor R&D combined to force national governments to assume control of their R&D programs in the initial stages. The exception is the Federal Republic of Germany. Although the surveyed nations differ in their emphases on these four factors, especially over time, they can be identified in each case. One might expect to find analogous institutions and organizational strategies dealing with nuclear energy in each nation. In fact, this is true, with only degrees of difference distinguishing the nations. The major exception again is the Federal Republic of Germany, a distinction largely explained by the German political structure and its lack of a weapons program. However, even with these differences, West Germany has slowly established R&D institutions and strategies analogous to those of the other nations.

---

[14] See Perry et al., *The Light Water Reactor*, for the U.S. experience.

These similarities can be grouped into the following areas: the centralization of bureaucratic authority, the creation of new and uniquely endowed government agencies, clear civilian control, close proximity and access to high levels of government decision-making, and a large degree of autonomy within the bureaucracy. Some important national differences also existed and will be noted.

Centralization of Bureaucratic Authority

The motivating elements discussed above centralized early nuclear research under government control. In each case, the central government moved—some more quickly than others—to assert its control even when its own political—if not bureaucratic—stability was uncertain. The postwar caretaker government in France under General Charles deGaulle created the *Commissariat à l'Energie Atomique* (CEA) on October 18, 1945, explaining that "Pressing necessities of a national and international order obligate taking the necessary measures in order that France can take her place in the domain of atomic energy research."[15] The United States, Great Britain, and Canada acted with similar alacrity to continue their centralized wartime control over nuclear matters. Of course, the Soviet Union offered no alternative to centralized governmental control of nuclear R&D. However, the Soviets also took measures to assure that nuclear research would remain the exclusive prerogative of the central government and would not be delegated to minor ministries or regional authorities. Their postwar research program was conducted under the leadership of "General Boris Lvovich Vannikov who in turn was strictly controlled by Lavrenty Beria, head of the secret police until his arrest on June 26, 1953."[16] Placing nuclear research under the directorship of Beria suggests the extreme importance attached to nuclear research by the Central Committee of Ministers. Even though one cannot distinguish expenditures

---

[15] The enacting Ordonnance No. 45-2563 du 18 Octobre is cited in Scheinman, *Atomic Energy Policy in France. . . .*, p. 7.

[16] Arnold Kramish, "Atomic Energy in the USSR," *Bulletin of the Atomic Scientists* 15, 8 (October 1959):323.

on Soviet weapons research from those on power reactor research, one can note that construction on the Obninsk reactor was begun in 1950 under this high level scrutiny.

In West Germany, the decentralized form of government, the primary responsibility of the individual states for scientific research, and the lingering fears of governmental control produced a different initial framework for nuclear research.[17] As Nau states, "Centralization was not unknown in German industrial history, but the Fascist experience had discredited policies of corporate concentration."[18] Therefore, West Germany was the only nation whose government did not quickly establish control over its budding nuclear research programs, but this was only a temporary reticence.

Creation of Special Government Agencies

Each national actor recognized that existing organizations and institutions would not be adequate for this new task, even in the American and British cases where weapons development was still to be the major concern of the research and in Germany where the government had longer to prepare for the institutional exigencies of nuclear power management.

In the United States, Great Britain, and Canada, strong arguments were presented that because nuclear energy would affect so many different areas of government, control should not be placed under one agency. Others argued that for the same reason, control of nuclear research must be centralized under one agency or ministry or risk the confusion of uncontrolled growth because of the absence of a coherent body to oversee it. In the United States, Secretary of the Interior Harold Ickes argued before the Senate committee assigned to prepare the Atomic Energy Act:

> Atomic fission is too big a matter to be set off in a field by itself with a fence around it and separated from the problems with which the regular departments of the Government have to deal. . . . In the space

---

[17] See Gerwin, *Atoms in Germany*, p. 13; and H.L.J. Burgess, "Nuclear Engineering in Germany," *Nuclear Engineering* 5, 46 (March 1960):114-115.

[18] Nau, *National Politics and International Technology*, pp. 85-86.

of our national existence the jump from man power to animal power to electricity in the field of production, communication and transportation was wholly revolutionary. Yet we did not find it necessary to set up a Federal department of steam or electricity....[19]

The opposing faction contended "that the development of atomic energy for the peaceful pursuits of mankind and for military purposes . . . should be directed and controlled by the federal government through a single permanent commission whose members . . . should be wholly devoted to that single duty and no other."[20] With the passage of the Atomic Energy Act of 1946, the U.S. government explicitly adopted the latter philosophy by transferring the wartime Manhattan Project from the War Department and creating an entirely new agency to control all forms of atomic energy and its research.[21]

The British experience was only marginally different. In November 1946, Parliament passed England's Atomic Energy Act, which transferred nuclear research from its wartime niche in the Department of Scientific and Industrial Research (where it had come under the innocuous heading of "Tube Alloys") to the Ministry of Supply, which was then Britain's defense procurement agency. This transfer reflected the primary concern of the Labour party's cabinet at the time for the development of nuclear weapons. It was not until the peacetime users of nuclear energy gained ascendancy in the mid-1950s that the British recognized that existing governmental entities were inadequate for handling the multiple ramifications of nuclear research. A government White Paper, *The Future Organization of the United Kingdom Atomic Energy Project*, concluded that "the most rapid and economical development in this field be secured by transferring responsibility from the Ministry of Supply to a nondepartmental organization."[22] Specifically, an entirely new organization, a public corporation "with no responsibility outside the field of

---

19 Senate testimony cited in Hodgetts, *Administering the Atom for Peace*, p. 10.
20 Ibid., p. 11.
21 The most complete documentation of this legislation is provided by Hewlett and Anderson, *The New World*, chs. 12-14.
22 Quoted in Hodgetts, *Administering the Atom for Peace*, p. 15.

atomic energy," was legislated in 1953. This was the United Kingdom Atomic Energy Authority (AEA).[23]

President deGaulle, encouraged by his scientific advisors, acted with customary decisiveness to form the French CEA.[24] Like its American and British counterparts, CEA also was an entirely new organization with no distracting responsibilities outside of conducting nuclear research. Its nominal research charter was concerned with nuclear energy.

Canada continued its postwar nuclear research program under the wartime administrative agency, the National Research Council. However, the Canadian government soon chose to diffuse its nuclear R&D and its authority with the formation of two Crown companies. The first, Eldorado Mining and Refining, Ltd., was created in 1944 and given a complete monopoly for supplying and refining uranium ore. The second agency, the Atomic Energy Control Board, was organized as part of the Canadian Atomic Energy Act of 1946. This was largely a regulatory body. By 1952, it was apparent that Canadian nuclear research was losing its organizational focus, and thus a third institution, Atomic Energy of Canada, Ltd. (AECL), was formed to replace the National Research Council as the operational agency for developing nuclear energy. With the exception of the regulatory purview retained by the Atomic Energy Control Board and the uranium prospecting and processing charter of Eldorado, AECL became the dominant Canadian agency with responsibility for nuclear research.

The West German government was unable or unwilling to assert federal governmental control during the initial period of nuclear reactor research in the late 1950s. Indeed, the German experience is the mirror image of the other nations considered. As Nau points out, "domestic and foreign policy considerations in Germany called for a cautious and fragmented approach to nuclear policies."[25] The federal government was initially

---

[23] See "The Structure of the Atomic Energy Authority," *Nuclear Engineering* 1, 1 (April 1956):20-21.

[24] Goldschmidt, *The Atomic Adventure*, writes that Joliot-Curie and Pierre Auger, another leading French scientist, convinced deGaulle of the efficacy of creating an organization specifically devoted to atomic energy research.

[25] Nau, *National Politics and International Technology*, p. 73.

pressured by the state government of North Rhine-Westphalia to develop German capabilities in nuclear research. In May 1955, the wartime restrictions imposed by the Allies were lifted, and in the same month, "a group of representatives from twenty electronic and chemical firms set up a study group to examine the new opportunities offered by nuclear research."[26] In response, the federal government established the *Bundesministerium für Atomfragen* (Ministry for Atomic Questions),[27] which assumed the responsibility for establishing policies delineating the areas of cooperation and competence among private, state, and federal authorities. The federal relationship between the central government (*Bund*) and the individual states (*Länder*) and the latter's autonomy in scientific research made the issue of *Bund* funding nuclear research and especially power reactors difficult to resolve. The *Bund-Länder* friction resulted in a federal nuclear research budget of approximately $11 million in 1956,[28] an amount hardly comparable to the other nuclear nation's research expenditures at analogous stages of development. Additional research funds were provided by the electric utility companies, the *Länder* governments, and, to a lesser extent, private industry. A governmental advisory committee, the *Deutsche Atomkommission* (DAK), composed of eminent scientists, officials, and industrialists, was created in early 1956.

Although the West German government had created special agencies for the management of nuclear energy, they were quite different from the British, American, and French counterparts in that they initially had very little authority. The states and the private industrial corporation were the early principal nuclear actors in the Federal Republic of Germany. One German official characterized the federal government's role as "very small bureaucracy, very large expert groups."[29] This condition led *Nuclear Engineering* to observe as late as 1960:

---

26 Ibid., p. 73.

27 In 1957, this ministry became the *Bundesministerium für Atomkernenergie und Wasserwirtschaft*, in 1962, the *Bundesministerium für Wissenschaftliche Forschung*, and, in 1969, the *Bundesministerium für Bildung und Wissenschaft*. For consistency's sake, this study refers to it as the Ministry for Atomic Energy (MAE).

28 Burgess, "Nuclear Engineering in Germany," p. 115.

29 Nau, *National Politics and International Technology*, p. 74.

> Of all the major industrial countries with atomic commitments, development work in Germany is probably the least well organized. The government is unable to finance construction projects although it can give assistance to research undertakings. As a result . . . there seems no way at the present time of raising sufficient finance for large-scale work to be undertaken. . . . There is a strong tradition in Germany for industrial research and development to be independent of any government interference and this, coupled with the independent attitude of the various *Länder*, results in a bit-piece form of development in which little cooperation between the various organizations can take place.[30]

By 1957, however, the federal government began to assume an increasingly active involvement. In 1957, DAK issued West Germany's first—albeit unofficial—five-year nuclear plan, the Eltville 500 MW(e) program, encouraged the nuclear program by seeking EURATOM subsidies, and urged private industries and electric utilities to become involved. A second, but still unofficial, five-year plan was announced in 1962. The first official five-year plan, enunciated in 1967, called for federal expenditures of over $1 billion, more than the two previous plans combined. The organizational trend in Germany for nuclear energy was thus replicating the trend observed in the other Western nuclear nations in centralizing government control over the nuclear power reactor research program through new agencies set up especially to manage nuclear R&D. This phenomenon is particularly striking in West Germany for two reasons lacking in the other nations. First, there was no national research program dedicated to building nuclear weapons systems whose charter could be expanded to include nuclear reactors, and second, the primary impetus to form the responsible federal agency came from organizations and institutions that were not part of the federal government.

The final nation is the Soviet Union. There has apparently been little movement toward the centralization of Soviet nuclear energy research organizations because all nuclear research was closely monitored and controlled by the top hierarchy of Soviet government from the very beginning. However, some evidence indicates a trend toward specialized agencies that has been observed in the other nations. Under Beria, the Soviet organization responsible for atomic energy research was known as "The First

---

30 "Germany," *Nuclear Engineering* 5, 55 (December 1960):543.

Main Directorate of the Council of Ministers of the USSR." After Beria's arrest and execution, the "atomic energy directorate was elevated to the status of a Ministry, i.e., the Ministry of Medium Machine Building."[31]

After becoming entangled in Soviet personality politics, a special sector of the ministry split off in 1956 to form the Chief Directorate attached to the Council of Ministers for the Utilization of Atomic Energy (*Glavatom*). Although it is difficult to ascertain exactly what the institutional purviews of *Glavatom* were—did it encompass both peaceful and military nuclear applications, R&D, regulation, and safety standards?—it is significant that Soviet leaders were sufficiently concerned with nuclear energy to create a special organization for it that reported directly to the Council of Ministers. If this were the case, the Soviet experience thus duplicated the Western experience creating special agencies whose sole responsibility is the management of nuclear energy research.

Civilian Control, Access, and Autonomy

Two other important commonalities can be observed among these nations. The first is the seeming requirement for civilian control of the new agencies created to control nuclear energy, even in those nations where military applications were central in the research programs. The second is the ready access of these agencies to the decisionmakers in their governments. A third characteristic, although not as universal or predominant, is the governmental autonomy that most of the research and development agencies were granted.

When it initiated its postwar nuclear research program, each of the nations was careful to specify and then select a civilian agency and nonmilitary personnel to lead its nuclear research. This was true even if the primary concern of the research was still dedicated to military purposes. The U.S. decision was typical. Two major actors urging President Truman to move the Manhattan Project out of the War Department and under civilian

[31] Kramish, "Atomic Energy in USSR," p. 323.

control were the military commanders of the project, General Leslie R. Groves and Secretary of War Stimson.[32] Similar sentiments were voiced in Great Britain. The military's participation was never broached in France, where the twin heads of CEA were chosen from the scientific community and among graduates of the elite *Ecole Polytechnique*, the traditional training facilitiy for French administrators. When CEA's military mission became more clear-cut, a military applications office was formed as part of CEA; no military personnel were placed at the top of the CEA hierarchy. In Canada, a military agency was never an issue. For obvious reasons, the Germans were careful not to impute any military tendencies to their nuclear program. The first minister for the German MAE, Franz Josef Strauss, was a leading politician in the Christian Democratic party.

The experience of the Soviet Union is less clear but generally supportive. General Vannikov (a civilian "soldier") remained with the nuclear program until after it was reorganized as the *Glavatom*; in late 1957, he was awarded the Order of Lenin, the highest state honor, "for services to the Soviet State and Soviet Army for the 'development of new techniques'."[33] Most of the Soviet atomic research administrators have been characterized by their technical backgrounds and their loyalty to the Communist party. Kramish wrote in 1959 that "the men who administer the Soviet atomic energy program characterize a unique breed of civil servants of the Soviet state. Developed over many decades, this high-level cadre of state servants is composed of hardened, devoted Communists combining technical, political, and administrative competence."[34] Furthermore, the Soviet leaders have rarely admitted professional military men to the highest councils and deliberations of government. Like the other nations surveyed, the Soviet Union chose to operate its nuclear research program under strict civilian control, even though it was largely devoted to weapons research and production during its early phases.

In every case, the nuclear research agencies have had close, im-

---

32 See Hewlett and Anderson, *The New World*, ch. 12.
33 Kramish, "Atomic Energy in the USSR," p. 323.
34 Ibid., p. 323.

mediate access to the decisionmaking centers of their governments, although during the tumults of the Fourth French Republic it was not always clear that a single political decisionmaker existed. This access was a necessity during the war years but was generally continued into the postwar period, as would be expected because of the close government supervision noted earlier of nuclear research. The United States established an executive agency that reported directly to the American president and to the cognizant committee in Congress. There were no intervening levels or other agencies. Similar channels of access were provided in France where the prime minister sat on CEA's governing board. The early ministers of the German MAE were formal members of the cabinet. The minister in charge of nuclear research in the Soviet Union sat on the Council of Ministers. Because of the separation of power between the state (the Council of Ministers and the party (the Politburo), membership on the former does not necessarily mean representation or influence on the latter. However, it is probably safe to assume that the nuclear minister had immediate access to the Politburo because of the importance the Soviets attached to nuclear research.

The British and Canadian cases are less obvious if viewed beyond the initial nuclear research periods. During the war years, the English and Canadian nuclear researchers had direct access to government policymakers because of the critical nature of their work. With the passage of the U.K. Atomic Energy Act in 1946, nuclear research was placed under the minister of supply; from 1953 to 1957, AEA reported to the lord president; and from 1957 to 1959, AEA reported directly to the prime minister. Since 1959, AEA has been part of the Ministry of Science whose minister represents AEA in the cabinet and Parliament. However, the Ministry of Science is dominated by AEA; furthermore, "the authority can, if it sees fit, deal directly with the minister or heads of the nationalized boards. The chairman of the authority attends meetings of the Cabinet Committee on Atomic Energy and has free access to the minister of Science and even the prime minister."[35] Even though AEA might be organizationally located in a ministry, clearly it is not politically isolated

[35] Hodgetts, *Administering the Atom for Peace*, p. 38.

from the relevant decisionmaking centers. The Canadian case is the most circuitous. "The resort to nondepartmental forms of organization has created in Canada ... an untidy situation in which the lines of authority and responsibility to Parliament and the executive are complex and somewhat devious."[36] AECL reports directly to the Privy Council Committee on Scientific Research, which was created to advise the government on scientific matters. However, this committee rarely meets; in effect, AECL therefore often reports directly to the cabinet. Still, the institutional framework does not provide the direct access observed in other nations.

A final institutional similarity among many of the countries is that their nuclear research agencies were generally accorded a great deal of autonomy and were consciously created to be outside the extant bureaucracies and their red tape. The purpose was to assure that the agencies would not become involved in everyday politics, or at least be as removed from the battle as any governmental agency could be. It was hoped autonomy would provide them with some freedom of action. This consideration was clearly apparent in the congressional testimony and legislation that led to the American AEC.[37] For example, AEC commissioners were appointed for five-year terms. Canada formed AECL as a Crown corporation—that is, not a part of the formal government structure. The French CEA was isolated from the French National Assembly and answered only to its governing board. Even though the prime minister (or his representatives) chairs that board, his vote was only one of eleven, although he appoints "three high civil servants." CEA was chartered such that the "Authority shall be administered ... in accordance with the general directions of the government."[38] However, the expertise of the general manager and the high commissioner usually dominated the governing body and directed the decisions of CEA unless the government had a clear and decisive preference.

The U.K. AEA was originally set up within the Ministry of

---

[36] Ibid., p. 50.

[37] See Hewlett and Anderson, *The New World*, ch. 14.

[38] Décret No. 51-7, January 4, 1951; cited in Scheinman, *Atomic Energy Policy in France ...*, p. 51. The 1951 decree was formulated, in fact, to reduce CEA's initial autonomy because of a Communist scandal within the agency. However,

Supply and was later reorganized outside the usual ministerial channels to report directly to the prime minister. It was later returned to a ministry, where it currently resides. Its organizational autonomy, then, has been a mixed experience, more dependent upon personalities than on the institution or the government's organization. The West German example was similarly mixed. The one case that is not mixed is the Soviet Union, which has permitted its nuclear research agency or laboratories no autonomy from the government's political or ideological dictates. There is some evidence from other Soviet scientific endeavors that this strict control has been deleterious, but confirmation from the nuclear field is not available.[39]

National Differences

Although there are many similarities among the agencies and institutions mandated to organize and carry out national research programs in nuclear energy, each set of national institutions has unique characteristics within its national context that distinguish it from other national programs. It is instructive to point out specifically what some of the more critical differences were that distinguished these agencies from each other and from others within their governments.

The U.S. Atomic Energy Act not only created an independent agency reporting to the executive branch, but it also created a special joint committee of Congress mandated to oversee its actions—the Joint Congressional Committee on Atomic Energy (JCAE). This dual responsibility, unique in American politics, assured Congress that the president would not have the sole voice in nuclear matters and that Congress itself could have a substantive role rather than merely reacting to executive initiatives. This emphasized the importance that the drafters of the legislation placed upon the future of nuclear energy as a weapons

---

even with this motivation, it is clear that the National Assembly intended to guarantee CEA some independence.

39 Leonid Vladimirov, *The Russian Space Bluff*, translated by David Floyd (New York: Dial Press, 1973), argues that centralized control for political and ideological purposes severely handicapped the Soviet space program.

and a power resource. It also indicated the importance of civilian control even when the primary research emphasis was military applications.[40] A second unique feature was AEC's internal structure of leadership and authority. Most American executive agencies have a single administrative head with extensive operational prerogatives; the Atomic Energy Act called for an administrative committee whose members were appointed by the president, subject to Senate confirmation, for five-year terms. As Hodgetts comments, this choice was not taken lightly:

> It was realized at the time that the choice of a commission "diffused responsibility" and would "slow down the decision process"; nevertheless, these defects were regarded as a "necessary sacrifice" in order to avoid concentration of responsibility in one person. Under the circumstances, it was thought better to entrust this untried field to a commission of equals, each member of which could bring his experience and judgment to bear on the issues involved.[41]

The watchdog function of JCAE reinforces this intention.[42]

The British AEA was also headed by a commission that remained intact as an institutional feature throughout AEA's organizational pilgrimage. Loosely fashioned after the example of the nationalized industries, it designated three existing nuclear establishments: basic research at Harwell, military development at Aldermaston, and industrial production at Risley (Lancashire). The heads of these organizations, along with Sir Donald Perrott, a career civil servant, and Lord Plowden, the chief economic planner in the Labour party, composed the first AEA commission. The commission was later expanded to fifteen members (this is a maximum number that need not be filled) and two additional functional areas, engineering and reactors. These additions reflect the shifting priorities within the U.K. nuclear establishment from military to civilian uses of atomic energy.

The French case is particularly interesting. Not only was CEA unique within France because of its generous allocation of

---

40 See Hewlett and Anderson, *The New World*, ch. 14.

41 Hodgetts, *Administering the Atom for Peace*, p. 32.

42 The best account of the AEC and the JCAE is Harold P. Green and Alan Rosenthal, (New York: Atherton Press, 1963), *Government of the Atom: The Integration of Powers.*

bureaucratic independence, but it was also unique in French bureaucracy and among the other nuclear nations in the division of its leadership between the scientific and the civil administration communities. This bifurcation is an attempt to reconcile the bureaucracy's need for authority and control over a scientific program with the scientist's need to pursue independent research.[43] CEA's charter explains its diarchic (or what Scheinman calls bicephalous) administration. It was originally organized to pursue "scientific and technical research with a view to the utilization of atomic energy in the several areas of science, industry and national defense."[44] Whereas other nationalized French industries—for example, electricity or coal—were headed by a civil servant, the scientific requirements of nuclear research made science an equal partner to administration. The most prestigious French nuclear scientist, Frédéric Joliot-Curie (at the time, the most recent French scientist to win a Nobel Prize), was able to couch the demands of the scientific community in politically imperative terms that President deGaulle could appreciate.

> I will state quite simply that if this country does not make the necessary effort to give science the importance it merits and to give those who serve it the prestige necessary for their influence to be felt, it will sooner or later become a colony.[45]

At the same time, it was unthinkable to permit the freethinking scientists who were difficult to control to administer a program of such vital importance to the nation. Therefore, CEA was formed with a high commissioner (a scientist) and an administrator-general (a civil servant). As Gilpin observed, "The large scientific-technical content of atomic energy as well as the importance and *mystique* of the subject called for an organization where scientists themselves shared influence with the tradi-

---

[43] This dichotomy is elaborated upon by Michel Crozier, *The Bureaucratic Phenomenon* (Chicago: University of Chicago Press, 1964), and is not isolated to nuclear R&D; for similar evidence in aviation R&D, see Alexander, "Weapons Acquisition in the Soviet Union, United States, and France."

[44] Ordonnance No. 45-2563, October 31, 1945; cited in Scheinman, *Atomic Energy Policy in France . . .*, p. 8.

[45] Quoted in Gilpin, *France in the Age of the Scientific State*, p. 151.

tional French administrative class, the *grande corps de l'État.*[46] Even though scientists had a significant influence in each of the countries studied for obvious reasons, nowhere was their influence as marked as it was in France. Part of this was due to the immense personal prestige of Joliot-Curie, who had fought with the French underground during the occupation as well as being a Nobel laureate. When he was forced to resign his post as high commissioner in 1951 because of his membership in the French Communist party, the institutional role of the scientists within CEA diminished as the Administrator-General, Francois Perrin, asserted and concretized his authority.

As initially created, the West German atomic energy agencies were characterized by their lack of authority. In this sense, they were almost the complete opposite of their counterpart agencies in other nations. As Nau states, "Altogether, the relatively informal and circumscribed authority of the Atomic Ministry stood in sharp contrast to the administrative power of the French CEA."[47] MAE's initial responsibility was only to establish guidelines, provide constrained funding limited to research projects, and coordinate state and private research programs; DAK was merely an advisory board. Not until the early 1960s did the government begin to recognize the need for increased governmental control; centralization tendencies by the federal government became evident in the mid-1960s and were manifested in 1967 with the publication of the first official five-year plan and German initiatives in large-scale R&D programs.

In Canada, nuclear research responsibilities that were vested in one agency in England, America, and France were split into three Crown corporations—Eldorado Mining and Refining, AECL, and the Atomic Energy Control Board. The two operating agencies, AECL and Eldorado, are the most relevant; the control board's main activities are regulatory.[48] Eldorado was incorporated in 1945 as a Crown corporation under the Dominion Companies Act (1934), which permits the Canadian government to own a portion

[46] Ibid., p. 165.

[47] Nau, *National Politics and International Technology*, p. 74.

[48] See "Atomic Energy Control Board," *Nuclear Engineering* 13, 145 (June 1968): p. 504, for details.

of the company without having it be a government agency. Eldorado operates upon a patent issued by the secretary of state and is a public corporation. The directors are given shares, but they are endorsed over to the appropriate minister. Eldorado was chartered to exercise a monopoly in exploration, mining, and refining of uranium ore. Even though it has since surrendered its monopolies in exploration and mining, it has retained its exclusivity in the refining of uranium ore. Because of its operations in the marketplace, Eldorado is not dependent upon the Canadian government for funding. AECL is also a Crown corporation, but it is almost completely financed out of government revenues. It is the major organizational actor in the Canadian nuclear power reactor program.[49]

Russian nuclear research has not warranted an entirely new organization, and most obviously, the nuclear agencies have not been granted any independence or autonomy from the bureaucratic, political, and ideological controls that characterize the Soviet system of government.

The Role of Private Industry

The discussion has not yet explicitly addressed the participation of two other major groups of actors. The first is private industry and electric utilities, or what, in conjunction with the national government, we have termed the technology delivery system. The second is an amalgam of other concerned participants, such as universities, labor unions, environmental groups, and organized scientists. One distinction among the governmental, private industry, and other group actors is that their roles and intensity of involvement change critically throughout the period of the development of LWR. For example, labor unions or environmental lobbyists might be a critical element at one point and they may be neutral five years before or after. This section addresses the earliest period of the atomic power reactor development,

[49] J.L. Gray, "AECL in Canada's Nuclear Power Programme," *Nuclear Engineering* 13, 145 (June 1968):499, the President of the AECL, wrote, "We are the agency . . . without which there would be no Canadian programme."

when the surveyed nations' nuclear agencies first began to turn serious research priorities to nuclear energy, especially as to how it might be utilized as a power reactor. For this reason, it was argued that private industry's role should be less than during the later periods characterized by the commercialization and dissemination of reactors. This was found to be the case.

The role of private industry varies greatly among the nations considered. In West Germany, private industry was instrumental in turning the government's attention to nuclear research. In France and England, private industry played an insignificant role, and of course, there was and is no private industry in the Soviet Union.

The most important and in-depth participation of private industry during the initial periods of nuclear R&D was found in the United States and Germany. Most of the U.S. economy and a large part of the political and research systems were predicated upon free enterprise and the role of private industry. Many private firms were actively involved in the wartime Manhattan Project; duPont built and managed—albeit reluctantly—the plutonium-producing piles at Hanford, and Monsanto Chemical Company operated Oak Ridge's Clinton labs toward the end of the war. American private industry also represented a large, irreplaceable pool of talent; the employment of this talent was imperative given the scope of the scientific and technical nature of the research. Thus, it was to be expected that the early U.S. efforts toward weapon developments and peacetime uses of atomic energy relied heavily upon the expertise and participation of private industry.

One of the first reactor developments, the Daniels pile, was to have been staffed by engineers and physicists from various industrial participants, such as Westinghouse, General Electric, Monsanto, and Dow Chemical, rather than government scientists at the Oak Ridge facilities. Westinghouse was the industry leader in designing nuclear propulsion reactors. In 1947, it completed a study on the heat transfer properties of water, and in 1948, it received an AEC contract to build and test a land-based pressurized water reactor. Ultimately, this reactor became the propulsion plant for the nuclear submarine. In 1947, General Electric also became involved in nuclear energy research. In exchange for assuming the operational responsibilities for the Hanford plutonium plant, GE received government assistance in constructing its nuclear research laboratories outside Schenectady. Clearly, and by the conscious decision of the U.S. government, private industry was an active and necessary participant in the early postwar American nuclear research. At this point, American electric utilities considered nuclear-generated electricity as too distant and too beset with technological and financial uncertainties for them to become actively involved in the initial stages of its development.

If the U.S. government nuclear research program was characterized by intimate industrial participation, the situation in the Federal Republic of Germany was almost incestuous. Private industry in West Germany was a motivating element that led the West German government to initiate and then enlarge its nuclear research program in the late 1950s. Before then, the nuclear research option had not been available to West German industry. Although the federal government provided some research money and established research centers at Karlsruhe and Jülich, German industry and utilities precipitated and maintained much of the nuclear research momentum during the early period. The first German reactor at Kahl, which became critical in November 1960, was a small—15MW(e)—BWR plant built by a German electrical construction company, the *Allgemeine Elektrizitätsgesellschaft* (AEG), which held the GE license in Germany. The first demonstration reactor at Gundremmingen, rated at 250 MW(e), was also a BWR built by AEG; it became critical in October 1966. Both of these plants were sponsored by two large German utilities, *Rheinisch-Westphälisches Elektrizitätswerk* (RWE) and *Atomkraftwerk Bayern* (AKB). Unlike the early U.S. demonstration power reactors, the German demonstration reactors were not financed by the federal government but by state utilities who "could make technical and economic assessments and obtain engineering experience."[50] Furthermore, the plants were constructed by private industry under contract to local users. This strategy was not coincidental. "Industrial circles, supported by the Economic Ministry under Lud-

[50] Burn, *The Political Economy of Nuclear Energy*, p. 105.

wig Erhard, opposed central planning in the nuclear sector and advocated the 'normalization' of this sector, that is, the integration of the nuclear activities into the market economy much like any other commercial undertaking.''[51] Not until the magnitude of the nuclear undertaking and its attendant financial and research commitments were fully realized did the decentralized nature of the German nuclear program begin to change.

Canada had less private involvement than the United States or West Germany, but there was still a greater amount of participation than was found in England or France. The incoproration of Eldorado Mining and Refining as a Crown corporation represented an element of private enterprise in the Canadian program. This was reinforced by the Financial Administration Act of 1951, which permitted Eldorado to own its own assets and formulate its own operating budget without annual appropriations from Parliament. The Canadian government also relaxed its control over nuclear energy when it relinquished Eldorado's monopoly on the exploration and mining of uranium ore. AECL is similarly a Crown corporation, but it is much more closely controlled by the government. The regional utilities were to play an important role in the development of the CANDU reactor system but not until a later period.

In France and England there was minimal participation by the private sector and power utilities during this early period for a number of reasons. First, of course, the governments of both countries retained strict monopolies over all nuclear research. Second, the early fears of power shortages were abating because of the stabilization of energy import requirements in the early 1950s; this also reduced the cost of conventional energy resources so there were fewer financial incentives for private industry. Third, both nations had nationalized their electric utility industries. The French *Electricité de France* (EdF) and the British Central Electricity Generating Board (CEGB) were both controlled by the state and therefore unable to act as independent participants in the development of the national nuclear programs.

The French example is representative of both nations. EdF

---

[51] Nau, *National Politics and International Technology*, p. 85.

held a legal monopoly on the production, transmission, and distribution of electricity in France. Its directors viewed nuclear power as too expensive and too uncertain to warrant investment by EdF, but its future was promising enough that they participated in a CEA-controlled advisory group, which brought together representatives of CEA, EdF, and industry representatives (the PEON group). The role of French industry was equally diffident. Nau has written that "in this early phase, French industries played a supportive role in the civilian program as suppliers of reactor materials and components. They exerted little influence in determing the overall direction of French policy."[52] The British experience was similar. The only substantial contribution to either nation's nuclear program by private industry was in the construction of the research facilities and reactors. In both nations, construction firms formed consortia that acted as industrial architects. Generally, these consortia were formed with the expectation that the nuclear power industry would later become a major growth industry, thus providing a profitable demand for their services.

In the Soviet Union, which lacks private industry, the control of the central government over all the relevant ministries obscures the role that the other segments of the Soviet industrial community might have played during the initial period of Soviet nuclear research. For example, it is difficult to ascertain how (or if) the Soviet heavy construction industry aided (or hindered) the nuclear R&D processes during this early stage. In the absence of evidence, the Soviet case in this aspect must remain blank for this period.

Lobbies and Nuclear Research

The final set of actors in this early stage were the various interested groups who were not in governmental and industrial categories, including labor unions, universities, atomic scientists, and so forth. During the early stages, by far the most important

[52] Ibid., p. 82.

were the nuclear scientists, especially as they acted in political capacities.

Both as individuals and in organizations, members of the scientific community had the greatest effect upon the American, British, and French nuclear programs. One obvious reason for their importance was that the nuclear programs could not progress without their professional training and intellectual skills. A second reason was that the public and their elected representatives tended to accept the scientists' expertise in this new area that had great uncertainties. Legislators were particularly responsive to scientific advice, a situation upon which the scientists were quick to capitalize.

The American wartime atomic bomb program owed an invaluable debt to its scientists. Nuclear physicists working in the United States first alerted President Franklin Roosevelt to the potential destructive capabilities of atomic energy in the famous letter signed by Albert Einstein in October 1939.[53] (Although working in the United States, such scientists as Fermi, Szilard, Teller, Einstein, and others could hardly be classified as "American scientists.") It was their activities that led to the formation of the Manhattan Project to develop the atomic bomb and their efforts that actually implemented that charter. Although the Manhattan Project was headed by an army general, the work of such scientists as Oppenheimer, Lawrence, and Fermi drove the project to its successful conclusion. Thus, the American scientists had a tremendous credit to draw upon when the United States turned to the peacetime developments of nuclear energy research.

This scientific debt was not long in being called due. A large number of scientists working in the various government laboratories were particularly concerned about their postwar freedom to conduct research; they wished to be released from the constraints of wartime security and classification. Thus, there was a general outrage against the security provisions of the May-Johnson bill, which the War Department submitted to Congress as the administration's proposal for the organization of the postwar nuclear energy. Two things were particularly offensive

---

[53] Hewlett and Anderson, *The New World*, pp. 16-17.

to these scientists. First, the May-Johnson bill gave the military an active role in formulating the postwar nuclear research agenda. The scientists objected on the grounds that it would severely restrict their intellectual freedom, especially for publication. Second, there were heavy penalties attached to the release of any information the armed forces might choose to classify. In spite of the attempts of the leading scientists to mollify their colleagues, opinions ran high. Herbert Anderson, who worked in the Chicago laboratories, wrote to a friend at Los Alamos:

> I must confess my confidence in our leaders, Oppenheimer, Lawrence, Compton, and Fermi, all members of the Scientific Panel, . . . who enjoined us to have faith in them and not influence this legislation, is shaken. I believe that these worthy men were duped—that they never had a chance to see this bill. Let us beware of any breach of our rights as men and citizens. The war is won, let us be free again![54]

Similar though less flamboyant sentiments were received from other members of the nuclear research community, much to the distress of the War Department and, in particular, General Groves, who feared that the release of classified information would provide the Soviets with the key to the atomic bomb. In late October 1945, the dissident scientists formed the Federation of Atomic Scientists (FAS). FAS was a persuasive ally in the congressional rejection of the May-Johnson bill and passage of the McMahon bill, which became the Atomic Energy Act of 1946. Largely because of FAS efforts, the military representation on AEC was excluded and the security provisions reduced. Although the legislation was the result of congressional politics, there is little doubt that U.S. atomic scientists had a decisive influence in both shaping the legislation and lobbying for its passage.[55]

The effect of American scientists on the early nuclear reactor research programs was not as obvious, although probably just as important. The nature of the atomic research community produced a multifaceted nuclear research program, a phenomenon best reflected by the large number of initial reactor research designs.

---

54 Ibid., p. 432.

55 The role of nuclear scientists in affecting the Atomic Energy Act is documented at length in chs. 12 and 14, ibid.

Rather than expend all the U.S. scientific resources on one type of reactor system, the freedom of research demanded and received by U.S. scientists permitted them to explore a number of research avenues.

The role of the French scientist in affecting his or her nation's nuclear research program was perhaps even more decisive in three ways. First, Joliot-Curie and his compatriots were accorded high level access and were not reticent in pressing their case.[56] As a result, not only did France become heavily involved in a nuclear research program even though her resources were extremely scarce, but a French scientist was also made the administrator of the program. Indeed, given the personal stature of Joliot-Curie, there was little question as to who was the de facto director of CEA.

Second, as Scheinman points out, "French scientists generally, and a substantial number of the Commissariat scientists in particular, inclined toward a statist economic philosophy."[57] This implies that they were especially concerned that private industry did not gain any undue influence over the development of nuclear energy, and they were successful in enforcing this exclusion. Third, they argued strongly for the proposition that the French nuclear research program should be dedicated to the peaceful uses of nuclear energy. The Commissariat was responsible for preparing the "peaceful purposes" statement made by M. Parodi to the United Nations in June 1946. It was not until the removal of Joliot-Curie and the concomitant decline of the influence of French scientists within the agency that CEA began to gravitate toward the military applications of nuclear research.

The attitude of the British atomic scientists was somewhat contradictory. On one hand, the scientific community argued for an independent British nuclear deterrent and, on the other, for the exclusion of the military from the research program's leadership. Regarding the first, an influential group of British nuclear scientists who had worked with the United States and Canada under the Quebec conference agreement were annoyed by the

---

[56] See Gilpin, *France in the Age of the Scientific State*, ch. 6.

[57] Scheinman, *Atomic Energy Policy in France . . .*, p. 34; also Gilpin, *France in the Age of the Scientific State*, ch. 9.

unilateral American actions foreclosing further sharing of nuclear technology.

> The fact was that the British had no lobby for their atomic interests in Washington. At this stage [late 1945] Britain's affairs as a whole were only of marginal importance to the Americans, and in atomic energy she counted for nothing provided she played the game with raw materials. . . . Even the American atomic scientists who had close and friendly relations with the British and who had fought for the passage of the McMahon Bill, did not interest themselves in its effects on Britain.[58]

The result was that many British scientists, perhaps somewhat out of pique, lobbied the government to establish a nuclear weapons program.

The leader of the Canadian wartime nuclear research program, Sir John Cockcroft, was recalled to England to become the Minister of Supply. Before he accepted the post, he received written assurance that "subject to rules for safeguarding military, the work of the establishment was to be carried on under conditions in which the exchange of view and publications of results would be similar to those in university research laboratories."[59] Thus, an initial fear of American nuclear scientists was addressed early in the British context. A second common fear, the role of the military, was also expressed, but to less avail. Lord Portal, the wartime Chief of the Air Staff, was appointed the Controller of Atomic Energy under Cockcroft. In short, the British nuclear scientists were concerned about their independence but not nearly so vocally or effectively as their American counterparts. An English Association of Atomic Scientists was formed: "It was a highly-respectable body . . . [but] it largely confined its activities to valuable but inevitably small-scale education of the public."[60] The one policy area in which British nuclear scientists did have influence was in their postwar encouragement of the government about both the feasiblity and desirability of a nuclear weapon program. In this respect, their contribution differed from that of the scientists in other nations.

---

58 Gowing, *Independence and Deterrence*, p. 111.
59 Ibid., p. 39.
60 Ibid., p. 50.

In Canada, scientists as a group seemed less forceful than elsewhere. They were most influential as individuals, but their main contributions were less as lobbyists and more as administrators. Eggleston accords high praise to Dr. C.J. MacKenzie, but less for his scientific capabilities than for his wartime administration of the National Research Council. The premier scientist during the wartime collaboration was British scientist Sir John Cockcroft, who was called back to England after the cessation of hostilities to direct the British nuclear research program.[61] The other major scientist involved in the Canadian reactor research program was also an Englishman and administrator, W. Bennett Lewis, who was chosen as Cockcroft's replacement. Generally ascribed the "principal credit for the CANDU concept,"[62] Lewis assumed Cockcroft's post at Canada's Chalk River Nuclear Laboratories in September 1946; in Eggleston's words, "the subsequent story of nuclear research in Canada will always bear the deep imprint of his personal attainments."[63] Still, there seems to have been little organized activity among scientists in Canada that significantly affected Canada's nuclear research strategy.

The situation in the Soviet Union is relatively difficult to evaluate. Soviet literature is replete with laudatory references to personal accomplishments and the vital contributions of Soviet scientists in theoretical nuclear physics, the war research effort, and early reactor research. The most famous Soviet nuclear scientist was Igor' Vasil'evich Kurchatov of Leningrad's Institute of Physics and Technology. He was a central figure in the first proposal of nuclear weapons (1940), in the design of Russia's first uranium-graphite reactor (December 1946), and the first nuclear explosion (November 1949).[64] Kurchatov's public stature was not an isolated incident; Petrosyants is careful to acknowledge the

---

[61] See Eggleston, *Canada's Nuclear Story*, chs. 4-9, for the wartime experience.

[62] This accolade is "generally acknowledged" according to Hugh C. McIntyre, "Natural-Uranium Heavy-Water Reactors," *Scientific American* 233, 4 (October 1975):21.

[63] Eggleston, *Canada's Nuclear Story*, p. 146.

[64] See V.S. Emelyanov, "Nuclear Energy in the Soviet Union," *Bulletin of the Atomic Scientists* 27, 9 (November 1971):49. At the time of the article, Emelyanov was the chief delegate of the USSR to the IAEA as well as serving on the State Committee on the Use of Atomic Energy.

contributions of numerous other individual Soviet scientists and research institutes, although he clearly accords Kurchatov a status of first among equals.[65] Even though the Soviet literature effusively describes Soviet scientists and their contributions to nuclear research, the complete domination by the Soviet state of its nuclear research during the early stages makes it difficult to imagine that Russian nuclear scientists could have voiced an independent policy about nuclear energy research beyond describing the theoretical knowledge and technical applications.

German science in the nineteenth century was widely decentralized with little central government coordination within the German university system. Gilpin writes, "The university, in effect, became the seat of government for developing scientific enterprise. . . . German universities were creatures of the individual states; they enjoyed a considerable degree of institutional autonomy."[66] Government efforts during both world wars to centralize or at least focus the disparate system toward military applications were not especially successful. Even the Nazi's directives to develop an atomic bomb were not fruitful.[67] This tradition was only the first obstacle that the postwar German scientific community had to overcome in order to engage in nuclear energy research. The second, and perhaps more formidable, obstacle was that nuclear research had been forbidden for the ten years following the war, thereby rendering German experience, scientists, applications, equipment, and techniques obsolete. A third obstacle was an implicit struggle between the older, somewhat discredited generation of German scientists and their younger colleagues. The former wished to revert to the traditional, decentralized mode that was characterized by institutional rivalries. The younger generation wished to avoid the debilitating effect that the traditional organizational structure might induce.

---

65 Petrosyants, *From Scientific Search to Atomic Industry*, p. 20. Petrosyants' description of Kurchatov merits repeating: "He was a very cheerful person. His dark, radiant eyes would look at one gaily and cunningly. When Igor' Vasil'evich was pleased, he would stroke his beard. His friends lovingly called him 'the Beard', and he gaily responded to this nickname."

66 Gilpin, *France in the Age of the Scientific State*, p. 113.

67 Efforts by Nazi Germany to develop the atomic bomb during World War II are

The younger German scientists adopted a three-pronged approach to facilitate nuclear research. First, they urged the government to establish nuclear research centers, arguing that the universities did not have the requisite facilities for modern nuclear research. Second, they realized that German industry would have to lead the way. In this perception, they differed significantly from their French compatriots, who had displayed a distinct antipathy toward private industry. Third, they recognized their late start but felt no compunction against basing the German nuclear program on foreign technologies. German scientists were successful in all these tasks. The federal government established small but important nuclear research centers at Karlsruhe and Jülich, and German industry did, in fact, make critical contributions to the advancement of the West German nuclear energy program. Although most German scientists had a great deal of previous nuclear research and technology from external sources upon which to build, they did work toward an indigenous nuclear industry.

Reflections and Implications

Before continuing, it is worthwhile to reflect upon the pressures that led to the early institutionalization of the various national nuclear research programs, the resulting institutions and organizations, and the effects that these forces and institutions may have had on the different reactors power programs. In virtually every case, significant definitions of the programs occurred during this early period that were to affect reactor development throughout the program.

The U.S. nuclear program was affected by a number of conflicting forces, some pressing for and others warning against centralization. In the first camp, the military requirements and the tremendous financial costs made centralized government control under the executive branch of the government a foregone conclusion. The latter forces were represented by Congress (which man-

---

described by David Irving, *The Virus House: The German Atomic Bomb Project* (New York: Simon and Schuster, 1968).

dated JCAE when AEC was created), the intimate involvement of private industry, and the American scientists (who lobbied extensively for a varied program essentially free of military restrictions). The resulting mixed system had tendencies toward both centralized and decentralized control. Combined with the financial resources and the advantage of the wartime research project, the U.S. government was institutionally ready and able to pursue the development of a number of different reactor types. This tendency was reinforced by the active research and participation of private industry.

The United Kingdom was affected by a few of the same forces as the United States (e.g., military uses), but with the important additional motivations of reaffirming national prestige and the perceived need for nuclear-generated electricity to bolster her diminishing power resources. These additional elements dictated that Britain could not enjoy the extended and expanded research horizons that postwar America did. The immediacy of the British situation thus acted to reduce the United Kingdom's nuclear search in both time and scope. She could not afford to point her power reactor program toward the "optimal" reactor; rather, she was constrained to adopt a more "satisficing" search and selection strategy than the United States.[68] Furthermore, the British decision to develop nuclear weapons prior to the commercial development of reactors foreclosed a set of reactor options. The constricting actions were manifested by the relevant organizations. AEA was a fairly closed agency that sought and received little assistance or advice from other interested organizations. The secrecy of weapon-oriented research made much of AEA's early work inaccessible to other government agencies or knowledgeable civilian groups that could have broadened AEA's research scope. The nationalized electricity utility, which would operate any nuclear generating station, was treated as little more than a customer of AEA's development program. Private industry was effectively excluded from nuclear research except as a

---

[68] The theory of organization search and choice is set forth by James G. March and Herbert A. Simon, *Organizations* (New York: John Wiley, 1958); an excellent application of this theory to the political process is found in Allison, *The Essence of Decision*, chs. 3 and 4.

supplier because the government strictly controlled all elements of the program. British atomic scientists, unlike their American or French colleagues, were content to let the government define and manage the nuclear research program, including their participation. The British parliamentary form of government reduced the ability of Parliament to oversee or modify AEA's decisions.[69]

These British institutional factors resulted in a very small number of important decisionmakers who, with little or no national debate, chose to acquire a plutonium-producing plant to support an independent English nuclear deterrent. More than any other, this choice severely restricted the British nuclear power alternatives. Specifically, it meant that the LWR class of atomic reactors would not be available.

In many respects, the French nuclear program during this period was closely analogous to the British. French decisionmakers opted for a nuclear research program as a means of reasserting France's position in the international community as well as supplementing her decreasing fossil fuel supplies. Similarly, France chose to remain independent of American assistance and its attendant restrictions and to build a plutonium-producing facility to support an indigenous nuclear weapons program. These choices were perceived in the closed nature of CEA. All of the factors combined to restrict the French choice of power reactors. There was an important additional actor in France, the French scientists, who, unlike their British compatriots, were not politically quiescent. Their advice prompted the deGaulle government to organize CEA in 1945. The tremendous influence they wielded was reflected in the specification of a French scientist as the coadministrator of the body in the founding edict. The statist philosophy of the French scientists led them to argue against private French industry playing any significant role in their

[69] The readiness of the government to exclude Parliament from questioning its nuclear reactor program can be noted as late as 1953; see "British Debates for Atomic Power," *Bulletin of the Atomic Scientists*, 1953. For example, when a member of Parliament asked for the comparative costs of generating electricity from nuclear power or coal, the Minister of Supply, Duncan Sandys, responded, "we have made certain calculations, but I should not like to state them publicly" (p. 141).

nuclear program during this period. However, their antiweapons sentiments faded with the dismissal of Joliot-Curie.

For somewhat different reasons, therefore, France and England organized their nuclear research programs around centralized institutions that minimized the number of participants and limited the possible research strategies. There were few people in positions of knowledge and authority who could criticize the reactor programs chosen, few who could argue for other alternatives, and many who could and did defend the vested nuclear interests. Therefore, from an organizational standpoint, it is not surprising that the French and British nuclear programs were restricted to a specific type of reactor almost from the beginning and even less surprising that the later decisions to consider, let alone adopt, another reactor system were made with the greatest reluctance.

The early Canadian decision regarding a nuclear research program differed in two important ways from those reviewed above. In Canada, a conscious decision was made not to develop nuclear weapons. In addition, whereas England and France were stimulated to initiate nuclear research because of a fear of reduced power resources, Canada was richly endowed with hydroelectric power, and of course, her industry had not suffered war damages. The lack of these pressures permitted Canada to start and operate a nuclear research program at an orderly pace and in an open atmosphere. Even though the lack of enriched uranium meant that Canada was restricted in its choice of reactor types, as were the French and British, the absence of political or economic pressures to proceed immediately into nuclear power development gave Canada an advantage over France and England. This more studied approach is perhaps symbolized by the fact that AECL was not organized until 1952, well after the Canadian government had decided to develop nuclear energy to generate electricity. Even though the Canadian program was centralized subject to the prerogatives of the central government and basically restricted to one reactor type, the slower pace of the program permitted greater nongovernment participation.

The late entry of West Germany into the nuclear energy research field provided her with a sizable base of knowledge upon which to build. This, combined with latent (often manifest) fears of German nationalism and centralization, fostered an extremely decentralized German nuclear research program. The program was supported largely by the individual provinces and private industry, with the federal government providing only limited research funds, coordination, and advisory assistance. In fact, at variance with the other examples, it was the German atomic scientists who argued during this period that the government should move to centralize the German nuclear program and at least to establish nuclear research centers. The contributions of private industry were influential, as were the requirements of the individual state utilities for additional power plants. From an organizational standpoint, this decentralized context encouraged a variety of development strategies and approaches. Thus, the early German organization of nuclear research was such that a number of reactors could be examined, including the LWR, because the United States had guaranteed the Germans a supply of enriched uranium.

Finally, the Soviet Union chose to develop its nuclear energy program under the strictest control of any nation. The dominant institutional characteristic of the Soviet program was its extremely centralized control. The success of this approach might have prompted the Soviet leaders to maintain their tight control, although one can confidently surmise that they required little prompting. It is suggested that the early Soviet interest in nuclear research was represented by the formation and monitoring of special agencies, their location in important ministries, their proximity to the critical decisionmakers, and the propaganda value attached to Russian nuclear developments. All of these presaged the importance that the Soviets would place on nuclear research.

This examination of the factors motivating the different nations to organize and engage in nuclear research and the institutions formed to carry out that mandate is necessary to appreciate the research strategies and choices that emerged during the formative period of the national research programs.

5

Early Research and Development Strategies

Introduction: Single vs. Multiple Research Options

Each of the nations discussed had made a fundamental decision to develop nuclear energy. Each turned part of its research resources to nuclear power reactors. In most cases, this was a continuation or extension of its nuclear weapons research. Each nation organized special agencies to sponsor and direct its research and made a number of special provisions. After this was accomplished, the nations turned to the development of nuclear weapons and power reactors. This section describes the initial choices about nuclear reactors, how they were dictated by prior decisions, and the effect these early decisions would have upon the later course of reactor development.

The most important difference among the national research strategies was whether a nation chose to develop multiple or competitive reactor technologies concurrently or concentrate its resources on a single reactor type. In retrospect, the choice was always clear, but at the time, neither strategy appeared dominant in the face of severe technological and operating uncertainties, especially given limited financial and personnel resources.

The primary advantage of pursuing multiple technologies or alternatives is that it minimizes the risk of failure; if one alternative failed to perform as expected, a second or third type would be readily available. Upon development, it might happen that the third option would be superior to the originally designated first or second choices. The principal advantage, then, is that several different technological approaches offer a higher numerical probability for overcoming technical problems than a single approach. In areas of great technical and economic uncertainties, this strategy is often preferred. The drawbacks, however, are potentially severe. By committing scarce research resources across a number of options, it is possible that no alternative will receive sufficient attention, that none will be sufficiently developed, and that the development of multiple options will be more expensive than if a limited number were explored. Sooner or later, the multiple option strategy must be winnowed to one or two alternatives. Even with only two options, the same arguments regarding the multiple strategy apply, but this curtails both the benefits and the disadvantages. The third option, which is now precluded, might, with additional research, have proven to be the best, but the remaining two options will have sufficient resources available to push their development.

The single technology R&D strategy permits the research establishment to concentrate its resources on one technology. There is no danger of multiple options being individually underfunded or of too much money being spent on too many alternatives, many of which might prove to be unproductive. This strategy is often employed for developments considered to be too expensive for multiple approaches or models; for example, the U.S. Air Force did not have competitive hardware models of the costly B-1 bombers but it did for the less expensive lightweight fighter and attack aircraft competitions. Some argue that the single option research strategy is less expensive than the multiple option strategy. Although the empirical evidence for this proposition is mixed,[1] it is almost certainly false if the wrong choice is made.

[1] See Coulam, *Illusions of Choice*, pp. 203-215.

An important tradeoff between technology and uncertainty is implicit in selecting between the research strategies. If there is minimum technological uncertainty, the single option strategy is preferable; there is little reason to expend funds exploring alternative technologies. The multiple options strategy is preferable for minimizing the risk if technological uncertainty is large.[2] Although the state of knowledge for nuclear reactors during their earlier days was largely characterized by substantial uncertainties, France, Britain, and Canada chose the single technology reactor alternative. Meanwhile, the United States, the Soviet Union, and West Germany developed multiple reactor technologies (and alternatives within the specific reactor types) before deciding on a subset of reactor types for larger demonstration projects.

Instead of making generalizations and then categorizing the various national programs under the specific generalizations, this chapter is organized by the programs of the individual nations. However, a broad distinction is made between nations that adopted the single technology R&D strategy as opposed to nations that adopted the multiple technology R&D strategy.

This chapter also distinguishes between research oriented toward nuclear weapons and research oriented toward the power reactor. Except for Canada, no nation has terminated its atomic weapons research. The military and peacetime nuclear research programs cannot be cleanly separated from an organizational standpoint; for example, in the United Kingdom, AEA controlled the development of nuclear weapons and power reactors. However, the primary attention of this research is the development of nuclear power reactors. Hence, it necessarily focuses upon the peacetime development of nuclear energy. Although the military-civilian separation is not always institutionally distinct, in most cases the nations involved were beginning in this period to move from their primary emphasis on weapons research to the peacetime uses of nuclear energy as they found their initial weapons motivations fulfilled.

In every case, the first priority of the cognizant nuclear agen-

---

[2] Perry et al., *Systems Acquisition Strategies*, elaborate upon these two basic research strategies in the context of aircraft and missile procurement.

cies was to acquire or build research facilities and to train nuclear technicians and scientists. For the Soviet Union and the United States, much of this process was the result of the wartime atomic weapons program. The British and Canadian research programs were also outgrowths of this experience but not on the scale of the American or Russian programs. France and Germany initiated their programs after the end of the war and therefore could not draw upon the wartime base, and thus their emphasis on the construction of training and research facilities as a first priority was even more pronounced.

In most cases, the research facilities were entirely new, as benefits a new technology. In the United States, the research plants at Livermore, Oak Ridge, Hanford, Argonne, and the National Reactor Testing Site in Idaho were built specifically for nuclear research; this does not include other facilities that were strictly oriented toward weapons, such as Los Alamos and Alamagordo. Britain expanded its wartime research facility at Harwell and Risely to include test and prototype reactors as well as building new ones. The French CEA acted quickly to set up facilities at Châtillon and Saclay, explicitly "to build equipment for research and training of personnel."[3] The Canadians developed the Chalk River laboratories to meet their limited needs. Even in Germany, where the government was reluctant to become actively involved, nuclear research and training centers were established and supported by the *Bund* with additional *Länder* subsidies at Jülich and Karlsruhe.[4] The German reactor development program compensated for its governmental reticence by relying on its industrial complexes and university system to establish other centers for theoretical research and the training of nuclear personnel and by integrating foreign technology and personnel into the system.

The Soviet Union utilized its wartime plants for the development of nuclear trained personnel and prototype reactors, but it also depended heavily on its prewar laboratories and institutes. Perhaps the foremost of these were the Leningrad Physical and

---

[3] Perrin, "Atomic Energy in France," p. 92; as the chief administrator of the CEA, the author made as well as reported the strategy.

[4] See Burgess, "Nuclear Engineering in Germany," p. 114.

Technological Institute (referred to as the "cradle of Soviet physics") and the Moscow Physical Institute of the USSR Academy of Sciences.[5] Following the war, the most important Soviet nuclear research institute was set up in Moscow under the direction of Kurchatov, the Institute of Atomic Energy.[6] After the fabrication of the Obninsk reactor, the city of Obninsk became a major complex of nuclear laboratories under the heading of the Institute of Physics of the Chief Directorate for the Utilization of Atomic Energy. In early 1946, construction was started on what was to become the Joint Institute for Nuclear Research. The institute, which housed an early Soviet cyclotron, was located at Ivankovo (renamed Dubna in 1956). In addition to these institutes, the Soviets also trained nuclear personnel in their universities, such as Moscow State and Leningrad State Universities, both of which had extensive nuclear research and training facilities.

The early attention to training and research facilities was necessary in every case. Nuclear science required certain experimental facilities as a basis for further research. Access to experimental reactors was particularly important. Such facilities were necessary to train nuclear technicians and to provide theorists with equipment to advance their knowledge. Especially during the early atomic R&D period, when much research was devoted to weapons systems, nations were loath to train foreigners in their nuclear facilities for fear of revealing nuclear secrets. Therefore, the construction of training and research facilities was a common priority expected and found in each nation's nuclear program.

Multiple Reactors Development Strategy

The United States[7]

AEC chose to develop a number of alternative reactor research programs during the development period of its reactor RD&D.

---

[5] Petrosyants, *From Scientific Search to Atomic Industry*, ch. 1.

[6] Kramish, "Atomic Energy in the USSR," p. 324. For greater detail, see Petrosyants, *From Scientific Search to Atomic Industry*, "Atomic Science and Technology Centers," ch. 10.

[7] The United States will receive less attention here than the other nations

The conscious choice of a multiple reactor technologies development strategy was determined by at least five factors covering prior research experience, institutional considerations, and technology constraints. First, the experience with multiple research approaches to the atomic bomb had proven successful. The Manhattan Project had pursued six alternative bomb development technologies. Although costly, they were highly successful in producing an atomic bomb.[8] The U-235 gas diffusion process, originally thought to be one of the higher risk technologies, produced one of the two bombs dropped on Japan. Thus, American nuclear decisionmakers had a relevant and positive experience with concurrently developing multiple research options.

Other factors also motivated the American nuclear administrators and researchers to adopt a multiple technology development strategy. Except for the Soviet Union, the United States was the only nation that had a secure source and ample supply of enriched uranium. This allowed the United States to develop a family of reactors—particularly the LWRs—that were unavailable to nations without a source of U-235. Third, the active and intimate involvement of American utilities, private industry, and universities increased the number of knowledgeable participants, thereby expanding the areas or avenues of research and reducing the possibility that everybody would choose to follow the same research path. The emerging role of private industry and the electric utilities was demonstrated in AEC's industrial participation program, which began in 1951. Four industry-utility joint study groups were formed by AEC to evaluate the technical and economic merits of dual purpose (plutonium/power) reactors.

Just as important as the evaluations was AEC's recognition that the early and close participation of the concerns ultimately responsible would be critical in the commercialization of nuclear

---

because the U.S. reactor R&D is already well documented. See Hewlett and Duncan, *Atomic Shield*; and Wendy Allen, *Nuclear Reactors for Generating Electricity: U.S. Development from 1946 to 1963* (Santa Monica, Calif.: The Rand Corporation, R-2116-NSF, July 1977).

[8] See Groves, *Now It Can Be Told.*

plants.[9] Fourth, the United States was already well endowed with energy resources, such as hydroelectricity, petroleum, and coal. There was no pressing incentive to develop nuclear power immediately as a necessary supplement to diminishing fuel reserves. AEC could be more studied—almost leisurely—in deciding which nuclear reactors offered the greatest operational promise and therefore warranted AEC support. This multiple option was reinforced by the high degree of technological uncertainties perceived in the development of nuclear reactors; in other words, no single approach was seen as being dominant. The result was that the U.S. nuclear R&D policymakers were not forced to make a reactor choice in the immediate future, thus assuring a more extended list of reactor possibilities and a more studied selection process. Finally, the U.S. Navy's program to develop a nuclear propulsion system provided an added impetus for pressurized water reactor research that might not have otherwise been emphasized. For these general reasons, the U.S. nuclear decisionmakers chose to develop a number of different reactor technologies; for example, work was begun on an experimental breeder reactor as early as 1949.[10]

When JCAE in 1954 requested that AEC outline its five-year reactor plan, AEC presented five different reactor designs, even though the PWR variant of the LWR had already been designated as the most promising reactor for development.[11] The preeminence of the PWR, because of the work of the U.S. Navy and Westinghouse in developing it for the nuclear submarine program, gave the PWR a decided R&D advantage. This preference was manifested by AEC's decision the following year to build a PWR reactor as the first commercial prototype power reactor.[12]

In response to the JCAE request, AEC presented five reactor types that it had experimentally developed and tested since the

---

[9] See Allen, *Nuclear Reactors for Generating Electricity*, ch. 3, for an elaboration of the scope and effect of the study groups.

[10] See Samuel Glasstone and Alexander Sesonske, *Nuclear Reactor Engineering* (Princeton: Von Nostrand Press, 1963), p. 754.

[11] Hewlett and Duncan, *Nuclear Navy*, pp. 228, 234.

[12] The basic design and operation of the major reactor types discussed in this study are outlined and illustrated in Appendix B.

late 1940s in AEC laboratories or under AEC contract.[13] This number reflects the multiple option strategy and the large amount of technical uncertainty that characterized this stage of reactor development. The reactor types were the PWR and boiling water reactor of the LWR genus, the homogeneous reactor, the sodium cooled, graphite moderated reactor, and the breeder reactor.[14] A JCAE subcommittee estimated that the PWR and the sodium cooled reactor held the greatest technical promise for large-scale development testing within the following two or three years; the BWR was seen as five years away; and the breeder and homogeneous reactors, although thought to be possibly the most efficient in terms of producing economically competitive electricity, were viewed as being greater than five years away. Of the five proposed reactor types, the PWR was considered to be the least promising to produce competitive electric power.

Besides the fact that the first three reactors were water cooled—probably reflecting the American mastery of water physics and technology for nuclear engineering—there were few shared features among these reactor types.[15] The advanced development of two LWR variations underpinned the variety of reactor types that AEC and the participating industries were able to choose among and pursue. Not until Westinghouse and General Electric chose to build commercial-sized reactors did the U.S. government restrict its sponsorship of power reactor research and development except for further breeder reactor work. Even without government support, American private industry continued to test and develop two LWR reactor types (the PWR and BWR), many variations on those two basic types (e.g., "superheated" steam), and other reactors as well (e.g., an organic liquid cooled, heavy water reactor).

---

[13] Joint Committee on Atomic Energy, *Report of the Subcommittee on Research and Development on the Five-Year Reactor Development Program Proposed by the AEC* (Washington, D.C.: March 1954), pp. 61-64.

[14] The technical distinctions among these reactor types may be found in Allen, *Nuclear Reactors for Generating Electricity*, chs. 2 and 4.

[15] Weinburg, "Power Reactors," is a valuable perspective because it was written at the time.

Soviet nuclear technicians and decisionmakers also chose to develop a number of reactor types before making a commitment as to which reactors would be advanced to the demonstration stage. According to Wilcyznski, Soviet nuclear scientists fabricated experimental homogeneous, BWR cooled and moderated, carbonic acid cooled, heavy water moderated, light water cooled and moderated, and PWR cooled and moderated test reactors.[16] From these they chose to develop further a subset that included a light water cooled and graphite moderated (LWGR) at Obninsk, a BWR (Melekess), graphite-boiling water reactors (Beloyarsk 1 and 2), and PWR reactors (Novovorenezh 1-4).[17] The major difference between the Soviet and the U.S. multiple technology R&D strategy is that the United States carried out most of its experimentation quite briefly, early in its nuclear development cycle, and the Soviets seemed to have prolonged theirs. The Obninsk reactor began operation in 1954, a channel-graphite reactor in Troitsk in 1958, the Soviet BWR in 1962, the Beloyarsk reactor in 1964, and the PWR in 1969, although a PWR had powered the Soviet icebreaker *Lenin*, launched in 1958.[18] During the same time, the United States had tested a similar number of reactor types and had basically decided on the BWR and PWR. It was apparently not until the late 1960s that the Soviets decided to standardize their nuclear construction on the PWR and the LWGR (also a channel-graphite) reactors.[19]

The Soviets had research assets in Eastern Europe not

---

[16] Jozef Wilcyznski, "Atomic Energy for Peaceful Purposes in the Warsaw Pact Countries," *Soviet Studies* 26, 4 (October 1974):597. The broad range of Soviet experimental reactors is confirmed by Petrosyants, *From Scientific Search to Atomic Industry*, p. 198.

[17] The broad range of Soviet reactor types is emphasized in ibid., and in Philip R. Pryde and Lucy T. Pryde, "Soviet Nuclear Power," *Environment* 16, 3 (April 1974):26-34. Also see "Nuclear Power in the U.S.S.R." *Scientific American* 219, 7 (July 1968):50.

[18] For details, see "*Lenin*—The Russian Ice-Breaker," *Nuclear Engineering* 3, 31 (October 1958):432-433.

[19] See "W.H. Zinn Evaluates Russian Reactor Progress"; and Gillette, "Nuclear Power in the U.S.S.R.: American Visitors Find Surprises."

available to the other nations in this survey.[20] The Eastern European nations were permitted limited nuclear research by the Soviets, although it was even more strictly monitored and controlled than domestic Soviet research because the Soviets did not want to risk having nuclear materials fall into Eastern European hands and being covertly used in bombs. There is evidence that the Soviets allowed Poland and Czechoslovakia to engage in heavy water reactor research, an area in which Soviet scientists apparently had little experience. However, Russian interference in and control over the program were so pervasive that the Poles and especially the Czechs ultimately abandoned their program, and the Eastern European satellite nations were forced to purchase Soviet designed reactor units.[21]

Besides the actual reactor location and size, little else is known about the early Soviet reactor program. Table 5-1 lists the Soviet reactors by name, type, and dates when the reactor was planned, built, and began operation. The first Soviet nuclear test reactor—a uranium and graphite model—became critical on December 25, 1946.[22] This was probably the Soviets' main experimental reactor during this period. The first power reactor was the demonstration plant at Obninsk, which went into operation in 1954 and was announced to the world at the Geneva Conference on the Peaceful Uses of Atomic Energy in 1955. Even though it only produced 5 MW(e), the Soviets proclaimed that the water cooled, graphite moderated reactor was the world's first nuclear electric power station.[23] The Soviets gained some propaganda benefits from this claim, but they state that the predominant value of the Obninsk station was "the experience gained in designing, constructing, and maintaining that station."[24] A larger, 100 MW(e) LWGR, built in Troitsk (Siberia), went critical in 1958.[25] Information about the Troitsk reactors has been scarce, probably because their primary function is thought to be the pro-

---

20 See Jaroslav G. Polach, "Nuclear Power in East Europe," *East Europe* (May 1968):3-12.

21 Wilcyznski, "Atomic Energy for Peaceful Purposes . . . ," p. 587.

22 Emelyanov, "Nuclear Energy in the Soviet Union," p. 40.

23 Pryde and Pryde, "Soviet Nuclear Power," p. 26.

24 Emelyanov, "Nuclear Energy in the Soviet Union," p. 41.

25 Pryde and Pryde, "Soviet Nuclear Power," p. 27.

duction of fissionable materials for nuclear weapons. However, it is known that the first of the six reactors—a 100 MW(e)—went into operation in 1958.[26] Because it was also of the LWGR type, one can probably assume that it was likewise a modification of the Obninsk reactor. Beloyarsk 1, which was started in 1958 and went into operation on April 27, 1964, reflected and built upon these experiences. Rated at 100 MW(e), the Beloyarsk design was fundamentally the same as the Obninsk reactors and probably the Troitsk, with the important modification of having a higher thermal efficiency achieved by superheating in the reactor core. This improved the reactor thermal efficiency from 17 percent in Obninsk to 33 percent in Beloyarsk.[27]

While the Beloyarsk and Troitsk reactors were becoming operational, Soviet decisionmakers began to construct a second reactor type, the pressured water reactor. Like the American PWR, the Novovoronezh-1 reactor was partially an outgrowth of research on a naval nuclear propulsion unit, in this case, the Russian icebreaker *Lenin*, but to a much lesser degree. Rather than build a small demonstration or experimental plant analogous to the LWGR at Obninsk, the Soviets moved aggressively and opened the first Novovoronezh PWR, rated at 265 MW(e), in 1964. This reactor was to be the prototype for the Soviet PWRs.

The Mekeless BWR became operational in October 1965. Admittedly experimental in nature, it had a thermal capacity of 150 MW and an electrical capacity only a third as great. The Soviet nuclear technicians seemingly recognized that the BWR design inherently simplified the connection of the nuclear reactor with the turbine generators; there appear to have been no particularly troublesome operational problems. However, Soviet nuclear planners apparently decided not to advance the BWR past the development to the demonstration stage. With the exception of the four 12 MW(e) BWRs at Bilibino, which is in a remote area of Siberia,[28] no further Soviet BWRs have been built. Petrosyants'

---

26 Ibid., p. 27.

27 Petrosyants, *From Scientific Search to Atomic Industry*, p. 69; also see Knorre, "Frontiers of Technology: USSR," p. 317.

28 The Bilibino reactors are described by Pryde and Pryde, "Soviet Nuclear Power," pp. 28-29; their technical characteristics are listed in Petrosyants, *From Scientific Search to Atomic Industry*, p. 152.

Table 5-1. Nuclear Power Reactors in the Soviet Union

Reactor Site	*Reactor Designation*	*Reactor Type*	*MW Electricity*	*Planned*	*Date Construction Started*	*Completion (Actual or Planned)*
Obninsk	AM-1	LWGR	5	1951	–	1954
Obninsk	BR-5	FBR	5			1959
Troitsk	Siberian	LWGR	6 × 100	1954	1954	1958-62
NS Lenin	VVER	PWR	3 × 90 MW(th)	–	1956	1959
Sverdlovsk	Beloyarsk-1	LWGR	100	1957	1958	1964
Voronezh	Novovoronezh-1	PWR	265	1958	1959	1964
Melekess	VK-50	BWR	50	1960	1961	1965
Sverdlovsk	Beloyarsk-2	LWGR	200	1959	1962	1967
Voronezh	Novovoronezh-2	PWR	365	1964	1964	1969
Melekess	BOR-60	FBR	12	1963	1965	1970
Shevchenko	BN-350	FBR	350	1963	1964	1973
Voronezh	Novovoronezh-3	PWR	440	1965	1966	1971
Vovornezh	Novovoronezh-4	PWR	440	1965	1969	1972
Sverdlovsk	Beloyarsk-3 (BN-600)	FBR	600	1957	1968	1977

Chukhotka	Bilibino	BWR	4 × 12	1965	1966	1973-74
Murmansk	Kolskaya-1	PWR	440	1966	1967	1973
Murmansk	Kolskaya-2	PWR	440	1966	1967	1975
Erevan	Armenskaya-1	PWR	440	1970	n.a.	1975
Erevan	Armenskaya-2	PWR	440	1970	n.a.	1975
West Ukraine	Ukraina-1A	PWR	1000	1971	n.a.	1976
Voronezh	Novovoronezh-5	PWR	1000	1969	n.a.	1978
Leningrad	LAES-1	LWGR	1000	1968	1969	1976
Leningrad	LAES-2	LWGR	1000	1968	1969	1976
Kursk	KAES-1	LWGR	1000	1968	1970	1976
Kursk	KAES-2	LWGR	1000	1968	1970	1976
Smolensk	Smolenskaya-1	LWGR	1000	1971	n.a.	1976
Smolensk	Smolenskaya-2	LWGR	1000	1971	n.a.	1976
Kiev	Chernobylskaya-1	LWGR	1000	1971	n.a.	1977
Kiev	Chernobylskaya-2	LWGR	1000	1971	n.a.	1977

n.a. = not available.

Source: Wilczynski, "Atomic Energy for Peaceful Purposes . . . ," p. 575; Pryde and Pryde, "Soviet Nuclear Power," p. 29; "Power Reactors '75," *Nuclear Engineering International* 29, 229 (April 1975), Supplement, p. 406; Gillette, "Nuclear Power in the U.S.S.R.," p. 1003.

account of the Melekess reactor prototype does not indicate the reasons for discontinuing the development of the BWR.[29]

Thus, from a large selection of experimental reactor types and three operating models, the Soviet nuclear decisionmakers chose to standardize and build two nuclear reactor prototypes, the LWGR and the PWR. (Experimental fast-breeder reactors were also being developed during this time, but they are not of concern here.) One possible explanation why the Soviets chose not to develop the Mekeless type BWR any further is that they decided to build only two demonstration reactors and that the PWR and LWGR models simply showed greater promise. Perhaps if the progress of the PWR or LWGR development reactors had been more troublesome or otherwise discouraging, the BWR might have received additional support and a large demonstration reactor would have been built. The primary advantages of the graphite moderated reactor seemed to be the similarity between it and the early reactors that produced plutonium, the relative ease of construction, and the putative greater safety of reactor operations.[30] The PWR apparently offered some construction savings and greater thermal efficiencies. Rather than reduce their option even further and choose between the two at the early stage of their development, the Soviet nuclear planners opted to continue their multiple development strategy, but it would now be restricted to two (perhaps three) alternatives. The reasoning was analogous to that in the United States; Soviet authorities apparently viewed reactor uncertainty as having been reduced to manageable proportions. At the same time, research and fabrication costs were rising. Petrosyants elaborates:

> The decrease in the number of thermal neutron reactor types being developed and constructed for nuclear power plants in the USSR is explained by the fact that the numerous suggested types of reactors have no decisive advantages as to their economic effect, i.e., in general they differ comparatively little in the economic and technological respect. At the same time, construction of many reactor types would be too hard for the industry, because a large structural and technological variety of equipment would be required. Meanwhile, the experience in machine-building technology indicates that the fewer the

---

29 The Mekeless BWR prototype is discussed in ibid., pp. 130-132.
30 Pryde and Pryde, "Soviet Nuclear Power," p. 27.

types of items or machines, the cheaper their manufacturing, and the higher their quality.[31]

Petrosyants claims that there were no financial or technological advantages among the experimental Soviet reactor types, which suggests scant technical criteria at this stage of development for the selection and early convergence upon the LWGR and the PWR types. However, the more complete burn-up of the uranium fuel and the ease of construction and safety elements probably singled out the LWGR; lower operation costs and the promised thermal efficiencies favored the PWR; and the BWR had little relative advantage to warrant its continued development. Between the two chosen reactors, the Soviets seemed to favor the LWGR, even though its design was inherently more bulky, and it had more complicated internal plumbing than the PWR.[32]

The reasons for the proliferation of Soviet reactor types probably mirror the American rationales, with the important addition that the Soviets have a long and noted development style characterized by multiple approaches.[33] This approach is occasionally observed in the United States, but is not nearly as dominant as it is in the Soviet Union. Otherwise, the extensive wartime research base, the careful, unrushed approach (even more so than followed in the United States) enhanced by an abundance of alternative energy resources, and the large number of research institutes produced a variety of different reactor types from which the Soviets could choose.

During the early period of Soviet and U.S. nuclear reactor developments, both nations experimented with a broad range of reactor types. One reason for this hallmark of the U.S. experience has been attributed to the active participation of U.S. private industry and universities, which assured the U.S. nuclear decisionmakers of a number of different research approaches and,

---

31 Petrosyants, *From Scientific Search to Atomic Industry*, p. 198. Also see Petrosyants, "Nuclear Energy—A Productive Force," p. 19.

32 This logic is assigned in retrospect to the Soviets by Gillette, "Nuclear Power in the U.S.S.R.," pp. 1004-1005.

33 For an elaboration of this thesis, see Arthur J. Alexander, *R&D in Soviet Aviation* (Santa Monica, Calif.: The Rand Corporation, R-589-PR, 1970); Perry, *Comparison of Soviet and U.S. Technology*; and Alexander, "Weapons Acquisition in the Soviet Union, United States, and France."

consequently, a number of different reactor types. The Soviet Union developed an equally large number of reactor types, but the Russian research was much more centrally controlled in its reactor developmental strategy, a situation that would seem more conducive to the single technology development strategy.

The closed nature of Soviet decisionmaking reduces one to careful guessing about the reason for the Soviet reactor spectrum. It is probably safe to speculate that the large number of Soviet nuclear research laboratories and institutes, combined with the curiosity usually attributed to the scientific mind, produced research tacks too diffuse and esoteric for the central government to control rigidly. Indeed, as Amann and his colleagues point out, "In 1962, at least as large a proportion of the total labour force was employed in [Soviet] R&D as in the United States, and the absolute number of persons employed in R&D was almost certainly substantially greater than in the United States."[34] Given the magnitude of the effort, it simply might have been impossible for the central government to ride close herd on the different research paths being explored, especially in the experimental stages.[35] In conjunction with the Soviet proclivity for multiple research approaches and the large amounts of uncertainty ascribed to nuclear reactor research, this resulted in a broader variety of nuclear reactor types than one might otherwise expect from a tightly centralized bureaucracy. Thus, as long as the programs did not appear disastrous or extraordinarily expensive, the Council of Ministers was likely to let each reactor's advocates have their way until a decision had to be made. This decision may have been forced by the needs of heavy industry to standardize only a few nuclear plants for economic purposes.

The Federal Republic of Germany

The final nation that examined a variety of reactor types before progressing to the demonstration phase was the Federal

---

[34] R. Amann, M.J. Berry, and R.W. Davis, "Science and Industry in the USSR," in E. Zaleski et al, eds., *Science Policy in the USSR* (Paris: Organisation for Economic Co-operation and Development, 1969), p. 385.

[35] Deutsch has suggested that government does not have the resources

Republic of Germany. As in the United States, this preference for a multiple R&D strategy can be partly explained by the active participation of German industry, universities, the individual *Länder*, and the regional utilities. Similarly, Germany enjoyed a guaranteed source of enriched uranium so its choice was not initially foreclosed by the reactor design restriction of neutron economy. The Germans had another advantage over the other nuclear reactor nations. The late entry of their nation into the reactor industry meant that most of the early reactor designs had already been examined and the more questionable ones discarded. As two German observers candidly admitted much later, "we were able to concentrate on the most promising nuclear technologies."[36] Finally, West Germany was able to draw upon and participate in the research experiences of its European neighbors; for example, EURATOM sponsored an organic moderated reactor research program (the ORGEL reactor) and OEEC's European Nuclear Energy Agency (ENEA) subsidized high temperature gas reactors projects in Germany and Britain (DRAGON).

This is not meant to suggest that the German MAE did not foster and encourage an indigenous research program. One of its first acts, in consultation with DAK, was the Eltville plan, which drew up a 500 MW(e) national power reactor program. "In this program, five reactors of different types were to be constructed coming into operation by 1965. . . . It was realized that if these reactors were to be built they would have to be based on designs from overseas, but it was foreseen that eventually Germany could evolve its own designs."[37] Each of the five experimental reactors was to have an output of 100 MW(e): a gas cooled, graphite moderated reactor; a heavy water cooled and moderated reactor (HWR); a LWR; an organic liquid cooled and moderated

---

necessary to monitor every occurrence in an advanced industrial society. Thus, multiple reactors might have been developed because the appropriate ministry did not have the resources or incentives to prevent or reduce such a proliferation. See Karl W. Deutsch, "Cracks in the Monolith: Possibilities and Patterns of Disintegration in Totalitarian Systems," in Carl J. Friedrich, ed., *Totalitarianism* (Cambridge, Mass.: Harvard University Press, 1954).

36 W.J. Schmidt-Küster and M. Popp, "The German Nuclear Programme," *Nuclear Engineering International* 18, 201 (February 1973):94.

37 See "Germany: A Review," *Nuclear Engineering* 7, 78 (November 1962):443.

reactor (OMR); and a high temperature gas cooled reactor (HTGR).[38]

Industrial study groups and sponsoring utility consortia were assembled to undertake the research efforts. German Babcock and Wilcox (B and W) and six Hanover utilities investigated the gas-graphite reactor. AEG and RWE were the primary advocates of the LWR in Germany. Arbeitgemeinschaft/Brown, Boveri and Cie/Krupp (ABK), in conjunction with fifteen utilities centered in Düsseldorf, were assigned to be the principal research agents for the German HTGR, the *Arbeitgemeinschaft Versuchreaktor* (AVR). Siemens and six Bavarian utilities were given the heavy water reactor research program. To conduct the OMR research, a new company, Interatom, was formed by the Atomic International Division of North American Aviation and the German firm DEMAG; they worked with a utility consortium based in Baden-Württemberg.[39] The Eltville plan particularly emphasized German research on the HTGR, a LWR whose thermal efficiency would be increased by nuclear superheating of the steam, and the HWR. "These choices manifested a clear concern for the national integrity and competitive place of German programs."[40] Not unexpectedly in light of the role of German industry and utilities, each of these reactor types had its own coalition of boosters; for example, AEG and RWE favored the LWR, and Siemens supported the HWR.

Even though German industry and MAE were dedicated to the development of an indigenous German reactor, the initial strategy adopted was twofold. B and W, Interatom, and especially AEG chose to maintain "a small staff to attend to problems arising locally, drawing on their extended experimental facilities abroad for major investigations and tending to offer designs thoroughly tested and explored in previous applications."[41] Siemens and ABK were "resolved, almost publicly pledged, to produce *German* reactors. . . . Siemens [was] known to be vastly

[38] Gerwin, *Atoms in Germany*, p. 29.

[39] "Germany: A Review," p. 443.

[40] Nau, *National Politics and International Technology*, p. 86.

[41] Burgess, "Nuclear Engineering in Germany," p. 115.

more interested in its own development work than in the facilities offered through its association with Westinghouse."[42]

All of the above industries were also careful to maintain and cultivate French, British, and American nuclear industries. This was partly to avoid the appearance of premature commercial nationalism, but primarily it was to provide Germans industry with access to previous nuclear research that could possibly be of great use should its own research efforts falter or appear to be lagging because of the late start. In the late 1950s, Siemens extended its licensing arrangements with Westinghouse and initiated talks with Britain's English Electric about gas-graphite reactors. AEG held the German license for GE's development of the BWR. General Atomics and Atomics International (North American) were represented in Germany by Gutenhoffnungshütte and DEMAG (Interatom), respectively, and Krupp collaborated with England's Atomic Power Construction.

Germany's R&D strategy reflected the general uncertainty about nuclear reactors during the late 1950s. It was by no means obvious which reactor design or system would prove commercially preferable. England's Calder Hall gas-graphite reactor had gone into operation in late 1956 to the general acclaim of nuclear engineers; the U.S. Shippingport PWR reactor did not become operational until 1958 and then it performed at lower efficiencies than expected. Nau cites the German science minister, Siegfried Balke, as explicitly acknowledging in 1957 the technological superiority of the gas cooled, graphite moderated systems that were being developed in France and England.[43] Countering this was the so-called Three Wise Men Report, *A Target for Euratom*, which had been commissioned by the EURATOM governments.[44] This report rated the U.S. LWR design as technologically equivalent to the European reactor designs. Thus, even though much of the uncertainty of nuclear theory and reactor design had been eliminated by the late 1950s, there was still sufficient uncertainty that West German nuclear R&D pro-

---

[42] Ibid., p. 115; emphasis in original.

[43] *Frankfurter Neue Presse*, April 11, 1957, cited in Nau, *National Politics and International Technology*, p. 131.

[44] Louis Armand, Franz Etzel, and Francesco Giordani, *A Target for Euratom*, May 1957.

ponents chose to pursue a multiple reactor development strategy. This choice was made more feasible because of the desire of German industries to develop some areas of nuclear technology to increase their profits, or at least prevent German utilities from purchasing foreign reactors. Table 5-2 lists the German nuclear power plants constructed during this period. An additional nineteen are planned for completion by 1984.

Of the five 100 MW(e) prototype power reactors envisioned by the Eltville plan, none was built according to the original plan or schedule. Still, three important prototype plants were built and research done on a fourth through EURATOM during the development period. In 1958, RWE contracted AEG to build a 15 MW(e) prototype BWR facility at Kahl am Main, Bavaria. It was turned over to RWE in late 1960,[45] and it went into operation seven months later, only three years after the order was placed.[46] The Kahl facility was built without any *Bund* financial support. In late 1960, MAE "came to the opinion that State support of reactor development and construction could no longer be avoided."[47] AEG, B and W, ABK, and Interatom were given 18 million German marks for reactor studies and Siemens was given 100 million marks for a multipurpose reactor (MZFR) that was to be built at the Karlsruhe research facility. The latter plant, a 50 MW(e), heavy water moderated and cooled reactor, was contracted for in May 1961; construction was begun in June 1962, and it became operational in 1965. The MZFR was the first power reactor developed entirely within Germany.[48] The estimated cost of the MZFR facility was 157 million marks, of which MAE paid 81 percent and the *Länder* and the utilities paid the remainder; Siemens' investment is not known.[49] The third reactor type, the HTGR, was built at the Jülich test facility. Rated at 15 MW(e) the AVR is estimated to have cost 50 million marks, with MAE, the utility consortium, and ABK dividing the costs, 56 percent,

---

45 R. Kühnel, "The 15 MW BWR at Kahl," *Nuclear Engineering* 6, 57 (February 1961):56-65.

46 See Gerwin, *Atoms in Germany*, p. 32.

47 "Germany," *Nuclear Engineering* 6, 66 (November 1961):474.

48 "MZFR," *Nuclear Engineering* 11, 116 (January 1966):28-29.

49 "Germany: A Review," p. 444.

40 percent, and 4 percent, respectively.[50] The AVR featured an advanced pebble bed technology and was completed in 1967. Additional German research on HTGR technology was conducted with the ENEA/British DRAGON test reactor. Finally, the projected OMR research was conducted as part of the EURATOM organic liquid test reactor, ORGEL, in Italy.

In December 1959, the German federal legislature passed the Atomic Energy Act. It stipulated federal licensing codes, established safety regulation procedures, and set a maximum financial liability by the state in case of an accident for a magnitude of risks larger than could be covered by the established insurance market.[51] Section 36 of the act set the figure at 500 million marks, or approximately $117.6 million. The bill permitted private ownership but specified government possession of fissile materials. "Declaredly, the Act [was] designed to allow the Government to promote and guide initiative in science, industry and trade, but, in no circumstance, to operate directly in the field."[52]

By the early 1960s, four distinct characteristics of the German nuclear research program had begun to emerge. First, German industry and utilities were clearly involved—indeed, motivating—Germany's nuclear research program during its very first stages in an integral manner. Second, a broad-based nuclear research program was devised, and although the envisioned five different 100 MW(e) prototype reactors were not built, four different models of atomic reactors were ultimately constructed. Third, German industry, although concerned about developing a domestic nuclear technology, showed no hesitation in seeking and reaching licensing arrangements with foreign (usually American) firms. And fourth, German nuclear development was uniquely characterized by the *Bund-Länder* relationship, which divided the public sector developmental costs during the 1950s as shown in Table 5-3; the individual *Länder* contribu-

---

50 Ibid., p. 444.

51 The text of the act is contained in Gerwin, *Atoms in Germany*, pp. 93-117; also see "The West German Programme Begins to Gather Momentum," *Nuclear Engineering* 4, 36 (March 1959):96, and "German Atomic Energy Act," *Nuclear Engineering* 5, 44 (January 1960):4.

52 Burgess, "Nuclear Engineering in Germany," p. 113.

Table 5-2. Nuclear Power Reactors in the Federal Republic of Germany

Plant Location	*Plant Name*	*Reactor Type*	*MW(e)*	*Date Operational*	*Primary Contractor*
Kahl, Bavaria	Kahl VAK	BWR	16	1962	AEG/GE
Leopoldshafen	Karlsruhe MZFR[a]	PHWR	58	1965	Siemens
Gundremmingen, Bavaria	Gundremmingen-1	BWR	250	December 1966	AEG/Ge
N. Rhein–Westfalen	Jülich AVR[b]	HTGR	15	December 1967	ABK
Lingen, Darme	Lingen	BWR	268	October 1968	AEG
Mosbach, Baden	Obrigheim	PWR	345	March 1969	Siemens
Kahl, Main	Grosswelzheim	BWR	25	July 1970	AEG
Beverung	Wurgassen	BWR	670	1972	KWU
Stade	Stade	PWR	662	May 1972	Siemens
Landshut	Niederaichbach	GCHWR[c]	106	1974	Siemens
Würms	Biblis A	PWR	1,200	October 1974	KWU/Hochtief
Brunsbuttel	Brunsbuttel	BWR	805	1975	KWU
Leopoldshafen	Karlsruhe	FBR	21	1976	Interatom
Philippsburg	Philippsburg	BWR	900	1976	KWU
Weckarwestheim	Neckar	PWR	805	February 1976	KWU
Würms	Biblis B	PWR	1,300	August 1976	KWU/Hochtief
Landshut	Isar Nieder	BWR	907	1976	KWU

Nordenham	Unterweser	PWR	1,300	December 1976	KWU
Üntrop	Üntrop THTR[d]	HTGR	307	1978	Consortium

[a]*Mehrzweckforschungsreaktor*, a German pressurized heavy water moderated reactor.

[b]*Arbeitsgemeinschaft Versuchsreaktor*, a German HTGR.

[c]Gas-cooled, heavy water moderated reactor.

[d]*Thorium Hochtemperatur-Reaktor*, a German HTGR utilizing Thorium.

Source: "Power Reactors '75," p. 396.

Table 5-3. Distribution of West German Public Funds on Atomic Energy, 1956-1961 (1,000 marks)

	Bundes		*Länder*		*Total*	
	DM	*Percent*	*DM*	*Percent*	*DM*	*Percent*
Research and Development	541,784.0	68.1	254,126.8	31.9	795,910.8	78.0
Reactor Technical Development	89,880.3	94.7	5,063.2	5.3	94,943.5	9.3
Training	21,893.3	87.4	3,155.4	12.6	25,048.7	2.5
Radiation Protection	15,803.0	91.5	1,469.5	8.5	17,272.5	1.7
Administration	58,483.0	66.9	28,936.1	33.1	87,419.1	8.5
Totals	727,843.6	71.3	292,751.0	28.7	1,020,594.6	100.0

Source: "Germany: A Review," p. 442.

tions are listed in Table 5-4. However, by the early 1960s, faced with growing development costs and economic uncertainties, MAE began to increase its support of both nuclear research and the construction and operation of nuclear generating stations in both absolute and relative terms. In 1962, then, "the 2nd German Atomic Programme, comprising mainly a Demonstration Nuclear Power Station Programme and one for the development of advanced test reactors" was announced.[53]

Single Reactor Development Strategy

France, the United Kingdom, and Canada all adopted a single reactor R&D strategy. One reason was the absence of a guaranteed or secure source of enriched uranium. This forced each nation to develop reactor types that were characterized by neutron economy. The construction and moderator materials had to have a very low neutron "capture" cross section—that is, one that would not absorb a large number of neutrons. Failing this, and lacking enriched uranium, the reactor could not maintain its chain reaction. Beyond this similarity, the national programs

Table 5-4. Contributions to Nuclear Energy by Länder, 1956-1961 (1,000 marks)

Länder	*DM*	*Percent*
Baden-Württemberg	53,568.6	18.3
Bayern (Bavaria)	25,748.1	8.8
Berlin (West)	30,594.1	10.4
Bremen	2,529.5	.9
Hamburg	17,279.1	5.9
Hessen	11,839.2	4.0
Niedersachsen	9,288.7	3.2
Nordrhein Westfalen	130,514.6	44.6
Rheinland-Pfalz	5,963.0	2.0
Saarland	–	
Schleswig-Holstein	5,426.1	1.8
Totals	292,751.0	99.9

Source: "Germany: A Review," p. 442.

[53] L. Memmel, "The Nuclear Industry of the F.R.G.," *Nuclear Engineering* 13, 148 (September 1968):748.

diverge. Canada based its reactor on heavy water as both the cooling and moderating medium. France and Britain both chose graphite as a moderator and air, later CO_2, as the primary coolant. The divergence was more than coincidental, for each nation was moved to its own reactor design program by a series of unique pressures described above.

Canada

The Canadian choice of heavy water cooled and moderated reactors was made very early, even before the end of the war, as it became clear that Canada would not be privy to U.S. nuclear technology. Drawing upon their wartime experience with heavy water and the abundance of native uranium ore, Canadian physicists built a series of small, experimental test reactors. The first, a Zero Energy Experimental Pile (ZEEP), became critical on September 5, 1945, making it the first nuclear reactor in the world outside the United States. The National Research-X-perimental (NRX) and National Research Universal (NRU) reactors followed in 1947 and 1957, the latter being rated at 200 MW(th). They were extensions of the ZEEP design philosophy. The NRU was the first reactor to be refueled in operation.[54] With the exception of a critical leak in the NRX's heavy water system in December 1952, when a power surge seriously damaged general fuel rods resulting in decontamination and reconstruction projects that shut down the facility for fourteen months,[55] Canada's reactor research program encountered few serious delays.

---

[54] G.C. Laurence, "NRU Operating," *Nuclear Engineering* 3, 22 (January 1958):19-25.

[55] See W. Bennett Lewis, "The Accident to the NRX Reactor on December 12, 1952" (Toronto: AECL Paper No. 232, 1953).

The radiation level within the reactor enclosure was so intense that clean-up crews could remain inside for only 90 seconds. One volunteer described the activity: "When it was our time to work, a team of three of us practiced several times on the mock-up to be sure we had the correct tools and knew exactly how to use them. Finally, outfitted with white protective clothes, we descended into the reactor and worked frantically for our allotted time.

"For several months afterward, we saved our feces and urine to have them

Beginning in 1951, Canadian nuclear engineers began to plan for a nuclear power plant,[56] and in 1953, AECL decided to erect a nuclear power demonstration (NPD) plant at Rolphton on the Ottawa River, about thirty miles downstream from the AECL's Chalk River Laboratories. Somewhat analogous to the U.S. AEC prototype PWR at Shippingport, the NPD was planned as a cooperative venture between AECL, industry, and a public utility. AECL was to provide its expertise; Ontario Hydro provided the site, the design, and construction cost of the conventional part of the facility, and it was to operate the plant once it was declared ready; and Canadian General Electric was to design and manufacture the components. Costs were to be split 70, 25, and 5 percent, respectively.[57] Its design was to be basically an extension of the NRX reactor, except that it would deliver 22 MW(e) into the power grid of Ontario Hydro.

However, before the design was finalized, AECL's Nuclear Power Plant Division staff at Chalk River proposed a number of substantial design changes that have since been incorporated into all the commercial Canadian reactors constituting the Canadian Deuterium Uranium, or CANDU, series. First, in place of a single large pressure vessel, it was proposed "that the fuel rods be enclosed in individual tubes through which the heavy-water coolant would flow at a pressure of 1,450 pounds per square inch. . . . [This] was perhaps the most important single decision in the development of the CANDU system."[58] Second, to reduce the neutron absorption potential presented by the multiple pressure tubes, the division's engineers suggested a zirconium (which has a very low neutron-capture cross section) alloy. To this end, Zircalloy-2, which had been developed for the U.S. nuclear submarine program, was introduced into the design. Third, it was

---

monitored for radioactivity. We had absorbed a year's maximum allowance of radiation in 1 minute and 29 seconds. There were no apparent after-effects from this exposure—just a lot of doubtful jokes among ourselves about death versus sterility." Jimmy Carter, *Why Not the Best?* (New York: Bantam, 1976), pp. 59-60.

56 See W. Bennett Lewis, "An Atomic Power Proposal," DR-18, AECL Paper No. 186, August 1951.

57 "Ontario Hydro's Nuclear Activities," *Nuclear Engineering* 13, 145 (June 1968):502.

58 McIntyre, "Natural-Uranium Heavy-Water Reactors," p. 22.

recommended that the reactor be rotated on its side so that the pressure tubes would be horizontal. This would facilitate fuel loading without having to shut down the reactor. At the same time, they argued that the fuel configuration should be discrete short bundles, which would increase the utilization or burn-up rate of the uranium fuel. With a single long rod of uranium, only the middle was completely utilized (to perhaps 10,000 MW/ton) and the ends might achieve only 40 percent of that. By inserting the fuel elements sequentially and pushing them through the central core until they were removed at the other end of the tube, the fuel would be irradiated relatively uniformly. Fifth, the insertions were to be bidirectional; that is, the fuel rods would be inserted and pushed through the reactor from opposite ends in alternate tubes to facilitate the more complete burning of the fuel elements.[59] A sixth and final change was the substitution of uranium oxide (UO_2) for metallic fuel. The NPD facility was completed in March 1962, became critical on April 11, and on June 4, "the first electricity produced by uranium fission energy was fed into the transmission lines of Ontario Hydro from the Rolphton station."[60] The NPD achieved its full 22 MW(e) capacity later that month.

The success of the NPD plant significantly shaped the future Canadian reactor program. The NPD reactor had admirably fulfilled its role as Canada's HWR prototype. Every reactor built after the NPD reactor incorporated its technology. The only variations have been in the use of other cooling mediums, such as boiling water and organic liquids. Table 5-5 lists the CANDU reactors. With the advances incorporated in the NPD facility, the next logical progression in the Canadian nuclear reactor RD&D program was a large-scale electricity generating demonstration reactor.

Great Britain

The British and French faced comparable situations. Both decided to develop their indigenous nuclear capabilities, to make

---

[59] "Power Reactors in Canada: NPD and CANDU Studied," *Nuclear Engineering* 3, 29 (August 1958):334-338.

[60] Eggleston, *Canada's Nuclear Story*, p. 332.

Table 5-5. **Canadian CANDU Nuclear Power Reactors**

Name or Location	*Type*	*Power Net MW(e)*	*Nuclear Designer*	*Utility*	*Operational Date*
NPD Rolphton	PHW	22	AECL and CGE	Ontario Hydro	1962
Douglas Point	PHW	208	AECL	Ontario Hydro	1967
KANUPP	PHW	125	CGE	Karachi Electric Supply Corp., W. Pakistan	1971
RAPP	PWH	203	AECL	Dept. of Atomic Energy, India	1972
RAPP 2	PHW	203	AECL	DAE, India	1976
Gentilly 1	BLW	250	AECL	Hydro-Quebec	1971
Pickering A	PHW	4 × 514	AECL	Ontario Hydro	1971-1973
Bruce A	PHW	4 × 750	AECL	Ontario Hydro	1976-1978
Gentilly 2	PHW	600	AECL	Hydro-Quebec	1979
Cordoba	PHW	600	AECL	Comision Nacional de Energia Atomica, Argentina	
Lepreau	PHW	600	AECL	New Brunswick Electric Power Commission	1979
Pickering B	PHW	4 × 514	AECL	Ontario Hydro	1980-1982
Bruce B[a]	PHW	4 × 750	AECL	Ontario Hydro	1982-1984
Darlington[a]	PHW	4 × 750	AECL	Ontario Hydro	1983-1985
Wolsung 1[b]	PHW	1 × 600	AECL	Korea Electric Co.	1981

[a]Planned.

[b]Under negotiation.

Source: Dr. G.A. Pon, *The Fundamentals of Nuclear Power* (Ottawa: Atomic Energy of Canada Ltd.), June 24, 1975, p. 12.

their own nuclear weapons, and to use nuclear energy as a way to regain their national position. Both had similar shared constraints: the absence of enriched uranium, the fear of an impending energy crisis, and a nationalized electricity distribution system. With this set of conditions and constraints, it is not surprising that the British and French independently arrived at analogous nuclear reactors during the early stages of their research programs.

The British civil nuclear power program was fundamentally shaped by England's military decisions to acquire nuclear weapons. Given the military need for plutonium, the first reactors were designed solely for the production of plutonium. "It was at first intended that these large reactors should be of a design similar to those at Hanford in the U.S., i.e., thermal (slow) neutron reactors, graphite-moderated and water-cooled."[61] Such reactors were to be fueled with natural uranium. Significant technical design problems soon arose. Water cooling permitted a satisfactorily high power density—hence, plutonium production rate—in the fuel, particularly if the exit water temperature remained low, as it could when there was no requirement for power production. However, light water is a voracious absorber of neutrons in a graphite moderated reactor. Therefore, in the late 1940s, British scientists cautioned that any loss of water could result in a serious plant accident. Sir Christopher Hinton commented on this safety problem in retrospect:

> The recognized instability of the Hanford reactors had led the Americans to locate them remotely in barren country. In large and thinly populated nations like the U.S. it is reasonably easy to choose sites having the necessary degree of isolation. In the United Kingdom, however, the problem of finding a site sufficiently remote to comply with the safety distances which had to be observed was extremely difficult.[62]

In the place of water, British scientists substituted air at atmospheric pressure as a coolant, with the explicit knowledge that such a system could not be used to produce electricity because the temperature of the air leaving the reactor would not be sufficiently high to create the steam necessary to drive electric turbines. Nevertheless, the priority need for plutonium was such that in 1947, the British began work on two Pu reactors at Windscale that were graphite moderated and air cooled. Completed in late 1950, the sole function of the Windscale facility was the production of plutonium. As Hinton noted, these were "cumbersome and extravagant plants, but they had the great merit of enabling us to produce plutonium at the date which had been demanded."[63]

The experience gained at Windscale led to the first British

---

[61] Hinton, "Atomic Power in Britain," p. 29.
[62] Ibid., p. 29.
[63] Ibid., p. 30.

power reactors. As they moved toward the development of a prototype power reactor, English scientists recognized the inadequacy of the Windscale reactor for power production. The reactor requirement for the efficient production of plutonium and electricity stood in contradiction. The air coolant exit temperature was approximately 220° C, far too low for power production. The higher temperatures necessary for generating electricity required the use of a gas coolant that would not absorb neutrons and would not react with graphite. Pressurized carbon dioxide (CO_2), which displayed both qualities, was chosen. Next, the fuel cladding, which had been made of an aluminum alloy, was not feasible for use at the new higher temperatures because of its limited heat tolerance and because at higher temperatures the aluminum fuel cladding material reacted with the uranium. To overcome this problem, the alloy Magnox (combining magnesium, beryllium, and calcium) was developed.[64] With natural uranium, the fuel materials had to remain metallic, and given the shifting emphasis in Britain to power production, higher fuel utilization rates had to be designed. Special heat treatments and alloying additions to the uranium metal fuel were developed to permit this. By 1953, much of this technology had been assimilated.

When the demand for additional plutonium became apparent, it was decided that a dual-purpose reactor be built for the production of both plutonium and electricity. The site chosen was across the River Calder from Windscale, Calder Hall.[65] As related above, in October 1956, Calder Hall electricity was royally switched into the distribution grid of the British Central Electricity Authority by Queen Elizabeth. With some justification, the British claimed that England was the world's leader in nuclear technology and the production of electricity from nuclear power.[66]

In February 1955, a government White Paper was issued enun-

---

[64] Whence is derived the name the Magnox reactor, which is the British graphite moderated, CO_2 cooled nuclear power reactor.

[65] See Kenneth Jay, *Calder Hall* (London: Methuen, 1961).

[66] The claim is cautiously advanced by ibid., p. 35, and at the British Nuclear Energy Conference in London the following November. The conference is reported in "Why British Claim Lead in Nuclear Power," *Nucleonics* 14, 12 (December 1956):S1-S12; and L. Rotherham, "Calder Hall Nuclear Power Station," *Science* 25, 3245 (March 8, 1957), pp. 452-454.

ciating the government's plan to construct twelve larger, power-only Magnox nuclear reactors by the end of 1965. These were based upon the Calder Hall design with the important addition of on-line refueling. The White Paper estimated that the "ten-year program would provide a capacity of about 1,000-2,000 megawatts. Thus nuclear stations would, by 1965, be producing one-fourth of the country's electricity, the equivalent to that produced by about 5 to 6 million tons of coal a year."[67] The capital costs, estimated to exceed 300 million pounds, and life expectancy data were extrapolated from the Calder Hall reactor. Finally, the government acted to include other organizations in the program. The White Paper stated:

> Responsibility for building the nuclear power system will be shared between private industry which will build the stations, the Electricity Authorities which will own and operate them, and the Atomic Energy Authority which will continue research and development of power reactors, will buy and fabricate the uranium fuel, and process the used fuel.[68]

To expedite the participation of heavy industry in what was obviously to be a very capital-intensive undertaking requiring critical technical skills, the British government encouraged manufacturers of conventional electricity-generating plants and equipment to form industrial consortia. "These combines could undertake the design and construction of the industrial nuclear stations required both in the British program and in the export market"[69] English industry hoped to capture. In 1954, the four consortia were formed: the Nuclear Power Plant Company; English Electric/Babcock & Wilcox/Taylor Woodrow (EE/BW/TW); General Electric Co./Simon Carves (GEC/SC); and Associated Electrical Industries/John Thompson Nuclear Energy Co. (AEI/JT). A fifth consortium member, Atomic Power Construction (APC), was organized two years later as a nonprofit corporation.[70]

---

67 This paper is excerpted by H.A., "The British Program of Nuclear Power," *Bulletin of the Atomic Scientists* 11, 6 (June 1955):225.

68 Ibid., p. 225.

69 Hinton, "Atomic Power in Britain," p. 31.

70 The industrial groups are described in "A Better Calder for the World

British nuclear planners were not deluded by the Magnox reactor. They recognized that its size and cost were substantial and its growth potential limited. For example, the Shippingport LWR pressure vessel was only one-eleventh the volume of the Calder Hall pressure vessel and had a larger rated electricity-generating capacity. Polach concluded that "the specific investment per kilowatt capacity is substantially higher for the Magnox-GCRs than for the light-water systems."[71] Rather, the Magnox reactors were seen in English circles as an interim measure to provide England with a means of overcoming its perceived energy crisis. Although the U.K. AEA hoped to export the Magnox reactors to help offset Britain's balance of payment difficulties, its main R&D efforts were directed toward the second and third generation of nuclear reactors.[72] The preferred alternatives were the advanced gas cooled reactors, which were thought to be the reactor for the 1970s, and the fast breeder, which was predicted to be operational in the 1980s.[73] To these ends, AEA authorized the construction of experimental test facilities for an advanced gas reactor (AGR) at Windscale and a fast breeder reactor at Dounreay. A complete list of British nuclear power reactors is found in Table 5-6.

Thus, the British had apparently adopted a circumspect R&D strategy for the development of efficient nuclear reactor technology. They concentrated their early research efforts on a reactor that was as sophisticated as their resource constraints could permit them yet well within their technological limits. Five industrial consortia were organized for the construction of the nuclear reactors. Steps were taken to develop a second and third generation nuclear reactor system. With this strategy as a guideline, AEA accepted tenders from the industrial consortia to build four Magnox reactors ranging in MW(e) capacity from 138

Market," *Nucleonics* 14, 12 (December 1956):S6-S9; also see "Structure of the British Nuclear Industry," *Nuclear Engineering* 8, 82 (March 1963):96-99.

71 Jaroslav G. Polach, "Nuclear Power in Europe at the Crossroads," *Bulletin of the Atomic Scientists* 25, 8 (October 1969):16.

72 See the Editorial, "Second Year of the U.K. AEA," *Nuclear Engineering* 1, 5 (August 1956):31. The Second AEA Annual report, reported on pp. 191-192 of this issue, affirms this strategy.

73 See "British Government Speeds Nuclear Program," *Nucleonics* 14, 12 (December 1956):S3-S5.

to 250. Frightened by the vulnerability of its Middle East oil supply, revealed during the Suez crisis, CEGB and AEA placed these orders between June 1956 and June 1957. In March 1957, the revised projected improvements in Magnox nuclear technology led the British to expand the stated goals in the White Paper of 1,000-2,000 MW(e) to a new goal of 5,000-6,000 MW(e) by 1965, an increase of at least 250 percent, without increasing the number of nuclear reactor stations.[74] This expansion was due to the increased efficiencies in nuclear power generation over the previous two years. When orders for reactors were received in 1959 from Japan and Italy, British hopes for profitably exporting the Magnox escalated. In short, the late 1950s were a period of nuclear energy euphoria in the United Kingdom. Most obvious indicators suggested that Britain's R&D strategy of focusing on one reactor type was to pay handsome dividends. However, as will shortly be seen, these dividends were to become serious debits.

France

The postwar French nuclear cupboard was even barer than England's. Although there was an important nucleus of trained atomic scientists, there were no wartime facilities to use as a base. Nor were the basic materials available for reactor research. The wartime allies had monopolized the world's supply of uranium. Had it not been for ten tons of uranium oxide hidden by the French in Morocco before the war, French scientists would have no uranium for their test reactors. As it was, this limited amount of uranium restricted the French to building a graphite moderated test reactor. Heavy water produced in Norway, which chose to honor France's prewar contract, was used as a coolant.[75] These two supply variables determined the composition of the first French reactor, the ZOE, a zero energy, sintered uranium oxide reactor that began operation in December 1948.[76]

---

[74] See Hinton, "Atomic Power in Britain," p. 33; also Schattman, "Britain's Nuclear Power Program," p. 1489.

[75] See Jules Gueron, "Atomic Energy in Continental Western Europe," *Bulletin of the Atomic Scientists* 26, 6 (June 1970):63.

[76] ZOE stands for *Zéro-énergie, oxyde d'uranium, eau lourde.*

From this position of nuclear poverty, CEA moved quickly to initiate a training program with the proper facilities being a first priority. An atomic research center was established at Châtillon, a few kilometers south of Paris, where ZOE was built. The CEA staff expanded from under 250 in 1946 to over 1,700 by 1971. Construction on a second research facility at Saclay, again south of Paris, began in July 1949. Saclay was the site of the second French reactor, EL 2, rated at 2 MW(th), and used CO_2 under a pressure of seven atmosphere as a cooling system. French nuclear scientists and administrators have since pointed out that this was the world's first gas cooled reactor.[77] A second priority of the newly formed CEA was to discover a secure source of uranium in both France and her overseas empire. By the end of 1950, both of these requirements had been successfully addressed. French nuclear decisionmakers were free to turn their attentions to designing and carrying out a French nuclear R&D strategy.

During what Scheinman calls the "formative years" of CEA,[78] the government's establishment of extensive training and research facilities is especially noteworthy in the context of a French society that had traditionally decentralized and under-subsidized scientific research. Equally important was the realization that the French government was now prepared to link science and industry and sponsor this relationship. Gilpin summarizes these events:

> The importance of Saclay . . . is that it symbolizes the beginnings of a transformation of French scientific institutions. In place of the small and poorly equipped laboratories of individual professors that had traditionally characterized French science, at Saclay limited scientific resources were consolidated in a complex of modern laboratories. In place of individualistic professors there were now teams of researchers and supporting technicians. The industrialization of basic research was begun.[79]

Once the *immobilism* produced by the firing of Joliot-Curie was resolved and a new high commissioner appointed, CEA turned to

---

77 For example, see Perrin, "Atomic Energy in France," p. 93, and Goldschmidt, "The French Atomic Energy Program," p. 39.

78 See Scheinman, *Atomic Energy Policy in France . . .*, Part 1.

79 Gilpin, *France in the Age of the Scientific State*, p. 169.

Table 5-6. Nuclear Power Reactors in the United Kingdom

Plant Location	*Plant Designation*	*Reactor Type*	*MW(e)*	*Operation Date*	*Primary Contractor*
Cumberland	Calder Hall	Magnox	138	October 1956	consortium
Dumfriesshire	Chapelcross	Magnox	240	March 1959	consortium
Gloucestershire	Berkeley	Magnox	325	November 1962	TNPG
Essex	Bradwell	Magnox	347	November 1962	TNPG
Cumberland	Windscale	AGR	40	February 1963	consortium
Caithness	Dounreay DFR	FBR	15	July 1963	JT
Ayrshire	Hunterston A	Magnox	360	1964	GEC/SC
Dorset	DRAGON	HTR	none	October 1964	OECD/ENEA
Wales	Trawsfynnyd	Magnox	580	March 1965	APC
Somerset	Hinkley Point A	Magnox	531	May 1965	EE/BW/TW
Gloucestershire	Oldbury	Magnox	547	April 1968	TNPG
Dorset	Winfrith	SGHWR	103	January 1968	consortium
Anglesey	Wylfa	Magnox	1000	July 1971	EE/BW/TW
Somerset	Hinkley Point B	AGR	1330	1974	TNPG
Caithness	Dounreay PFR	FBR	250	1974	TNPG
Aryshire	Hunterston B	AGR	1330	1974	TNPG
Kent	Dungeness B	AGR	1280	1976	APC

Durham	Hartlepool	AGR	1332	1978	BNDC
Lancashire	Heysham	AGR	1332	1978	BNDC
Essex	Sizewell B	SGHWR	n.a.	1981	n.a.
East Lothian	Torness	SGHWR	n.a.	1981	n.a.
Essex	Sizewell C	SGHWR	n.a.	1981	n.a.

n.a. = not available.

Source: "Power Reactors '75," p. 400.

defining France's First Five-Year Plan, which was presented to the French National Assembly in late 1951. The presentation emphasized the peaceful uses of atomic energy and the opportunity for France to supplement its energy resources with atomic energy. M. Felix Gaillard, then Secretary of State for Atomic Energy, stated,

> Several years ago France solemnly declared that she would not orient her efforts towards the fabrication of bombs. But she intends to give the greatest possible extension to fundamental research and to the practical applications of nuclear physics. . . . France must undertake the construction of large atomic reactors producing plutonium without delay . . . mainly for the production of electrical energy of which we have so great a need.[80]

CEA officials argued that France's national prestige would be grievously undercut unless nuclear power were developed:

> We must be aware . . . that if . . . these efforts can be reconverted toward industrial uses, there is no doubt that in a few years the energy sources put at the disposition of people would so profoundly and radically transform and augment their economic activity that the nations which do not have it will appear as helpless as the most backward nations of the world today appear in the face of modern nations.[81]

The National Assembly approved the Five-Year Plan without any serious dissent, and CEA immediately announced plans for a plutonium experimental reactor at Marcoule, G1. In May 1956, the facility went into operation. Marcoule G1 was graphite moderated, partially because of the scarcity of heavy water, but mostly because of graphite's greater efficiency as a plutonium producer. It was cooled by air at atmospheric pressure and fueled with Magnox-clad uranium metal. Rated at 40 MW(th) and 5 MW(e), it was a dual-purpose reactor but its electricity-generating capabilities were never operated; G1 was patently a plutonium-producing facility. The initial conditions that drove the British to build the Windscale reactors (e.g., lack of unenriched uranium and the desire for nuclear weapons) were also

---

[80] Quoted in Scheinman, *Atomic Energy Policy in France. . . .*, p. 69.
[81] Quoted from National Assembly debates, ibid., p. 75.

prevalent in France, although more covertly for nuclear weapons. Thus, it is not surprising that the Marcoule G1 and Windscale reactors were similar in intention and design. Table 5-7 summarizes the nuclear power reactors in France. Six more 957 MW(e) PWR are scheduled to be built by 1981, including Tricastin 2 and 3 (Pierrelatte), Dampierre 1 and 2 (Dampierre), Gravelines B3 and B4 (Dunkerque), and a 1000 MW(e) BWR at Orléans (St. Laurent 4).

Marcoule G2 and G3, which were identical reactors rated at 40 MW(e), became operational in 1959 and 1960.[82] The major difference between G1 and these was that the French scientists, like their British colleagues, substituted CO_2 under pressure as the coolant because CO_2 was able to function at the high exit temperatures required for efficient power production. Other significant changes were also incorporated. "Marcoule G2 and G3 were . . . the first in the world to use a pressure vessel and leak-tight containment. This, along with horizontal fuel channels and on-power fuel loading, distinguished them from the British Calder Hall machines."[83]

Between Marcoule G1 and G2, three important events occurred that directly affected France's nuclear reactor program and its R&D strategy. First, the national debates on the European Defense Community (1954) and the Euratom Treaty (1956) clearly revealed the French intention to manufacture nuclear weapons because of her refusal to ratify a treaty that prohibited them. The first indication of this decision was the establishment in 1955 of the *Bureau d'Études Général* (BEG) in CEA whose function was secretly to develop atomic weapons.[84] The political humiliation of Suez in 1956 solidified French resolve. The formal announcement of the French government to build an atom bomb, one of the last official acts of the Fourth Republic, was thus little more than a formality. This decision meant that the French nuclear reactor

82 See "The Marcoule Reactor G-2," *Nuclear Engineering* 4, 43 (December 1959):435-437, for details.

83 Gueron, "Atomic Energy in Continental Western Europe," p. 63.

84 The circumstances surrounding the creation of the BEG and the decision of the changing governments of the Fourth Republic to acquire a nuclear weapons capability is best described in Scheinman, *Atomic Energy Policy in France . . .*, Part 2.

Table 5-7. Nuclear Power Reactors in France

Plant Location	*Plant Designation*	*Reactor Type*	*MW(e)*	*Operation Date*	*Primary Contractor*
Bagnols-Sur-Céze	Marcoule G1	air-graph.	3	May 1956	consortium
Bagnols-Sur-Céze	Marcoule G2,3	gas-graph.	2 × 40	April 1959 May 1960	SACM
Avoine	Chinon 1[a]	gas-graph.	70	February 1964	consortium
Avoine	Chinon 2	gas-graph.	240	February 1965	consrotium
Ardennes	Chooz, SENA	PWR	319	April 1967	consortium
Cadarche	Rapsodie	FBR	0	August 1967	consortium
Avoine	Chinon 3	gas-graph.	500	August 1967	consortium
Brennilis	Monts Arrée (EL4)	GCHWR	77	October 1967	CEA/EdF
Orléans	St. Laurent 1	gas-graph.	500	March 1969	consortium
Orléans	St. Laurent 2	gas-graph.	530	August 1971	consortium
Lyon	Bugey 1	gas-graph.	560	1972	consortium
Avignon	Marcoule Phénix	FBR	250	1974	consortium
Fessenheim	Fessenheim 1,2	PWR	2 × 930	1976	Framatome
Lyon	Bugey 2,3,4,5	PWR	4 × 957	1976-78	Framatome
Dunkerque	Gravelines B1, B2	PWR	2 × 957	1979	Framatome
Pierrelatte	Tricastin 1	PWR	957	1979	Framatome
Orléans	St. Laurent 3	BWR	1000	1979	consortium

[a]Chinon 1, 2, and 3 are sometimes referenced as EdF1, EdF2, and EdF3.

Source: "Power Reactors '75," p. 394.

program during the initial period would be a stepchild to the nuclear weapons program within CEA. Indeed, Marcoule G3 was supported in the National Assembly debates by the Ministry of Defense and presented as necessary for the "National Defense."

Second, beginning in 1955, the nationalized French electric utility, EdF, began to assert itself in matters of nuclear power. Before this time, it had contented itself with its participation in the *Commission Consultative pour la Production d'Éléctricité d'Origine Nucleaire* (PEON), which was an advisory board to CEA composed of representatives of the government, industry, and EdF. In mid-1955, PEON approved two nuclear plants to be built at Avoine, Chinon 1 and 2. The reactor construction would be supervised by CEA, but for the first time, they would be operated by EdF. Chinon 1 was rated at 70 MW(e) and began operation in 1962. Chinon 2, much larger at 240 MW(e), became critical in February 1965. At the time the reactors were ordered, EdF "expressed no preference between gas graphite and American light water technique."[85] However, heavy pressures favoring the Gallic designs as the *filière prioritaire* were brought to bear by CEA, and both reactors ordered were gas-graphite. At the same time, however, there were indications that EdF, as the owner and operator of the French electricity distribution system and the new nuclear plants, would not be content to risk its nuclear future on only one technology, especially in the face of significant financial and technological uncertainties.[86] This became apparent in 1960 when EdF decided to construct a PWR in conjunction with the Belgian government, which held a Westinghouse license. The SENA facility was built in the Ardennes and rated at 319 MW(e). In June 1962, CEA announced that it would build a heavy water research reactor and power reactor, EL4, at Mont d'Arrée (in Brittany), rated at 77 MW(e).[87]

---

[85] Nau, *National Politics and International Technology*, p. 82.

[86] The possibilities of adopting the American LWR and the Canadian HWR technologies are briefly discussed in the Commission's 1964 reports; see *Rapports de le Commission Consultative Pour La P:roduction D'Electricité D'Origine Nucleaire, 1964-1969: Les Dossiers De l'Energie*, Ministere de L'Industrie et de la Researche, Paris, approx. 1970, Tome I, p. 14. Hereafter referenced as *PEON Reports*.

[87] R. Carle and R. Gibrat, "EL4 and the French Heavy Water/Gas Line," *Nuclear Engineering* 10, 108 (May 1965):171-174.

Significantly, EdF had the primary research responsibilities. During this period, the important organizational phenomenon was the beginning of the transition to EdF from CEA as the dominant French nuclear reactor actor.

Finally, CEA, EdF, and French industry began to recognize the growing magnitude of nuclear construction. Construction and design consortia, such as Indatom, Framatome, and GAAA, were formed. But, as Nau argues,

> These designs consortia, while acquiring considerable technical experience and expertise from CEA contracts, did not integrate manufacturing activities. They did not face, therefore, the critical investment choices associated with overall financial and economic responsibility which might have stimulated concentration.[88]

EdF was even less willing to include industry than CEA. EdF acted as its own industrial architect for the Chinon reactors and restricted industrial participation to merely that of supplying components. Thus, French industry was neither an integral nor knowledgeable partner in either proposing alternative nuclear designs or, perhaps even more important, in devising construction and operation economics so that the gas-graphite reactor could become an economically attractive reactor system relative to other nuclear reactors or fossil-fuel plants.

CEA and EdF did encourage the formation of professional associations, such as the *Association Technique pour le production et l'utilisation de l'Énergie Nucléaire* (ATEN), which was founded in 1955 "under the auspices of the CEA and EdF, to provide a central information service for its associates and for the public, and in this capacity to organize and coordinate participation at conferences, congresses and exhibitions."[89] Other organizations included the *Groupe Intersyndical de l'Industrie Nucléaire* (GIIN), founded in 1959, and BREVATOME, which controls the nuclear patents of its members, including CEA and EdF.[90]

---

88 Nau, *National Politics and Internal Technology*, p. 82.

89 M.H. Ziegler, "The French Nuclear Industry," *Nuclear Engineering* 5, 55 (December 1960):547.

90 "France: Industry—Its Organization and Associations," *Nuclear Engineering* 8, 80 (September 1963):305-306.

The French program moved rather cautiously ahead with its gas-graphite program during the early 1960s. With the exception of the Rapsodie FBR and the EL4 heavy water reactor, CEA preferred to concentrate its research resources on the gas-graphite reactor.[91] This emphasis plus the improvement of the economic conditions slackened the pace of French reactor R&D.

Three phenomena during this period were to have fundamental effects on the French nuclear reactor program within the decade. First, for the reasons discussed above, the French CEA chose to focus its R&D efforts on only one reactor type, the gas-graphite, and to preclude all others from their developmental purview. This strategy diverged sharply from what was seen in Britain, where the U.K. AEA was careful to recognize the limits of the gas-graphite nuclear plant and to acknowledge that it was only a first generation reactor with the advanced gas and the fast breeder reactors still to come. Second, EdF emerged as an important actor. It assumed a different strategy from its British counterpart, the Central Electricity Administration (later the Central Electricity Generating Board) and at an earlier stage. Whereas the British nationalized utility had made a monumental commitment to nuclear power and Magnox reactor, EdF adopted a more cautious, incremental approach. Finally, French industry had been largely excluded from the nuclear R&D, even more so than in England. This is in marked contrast to the situation seen in Germany roughly a decade later, or in the United States a decade earlier.

To summarize the French experience, the French R&D program was in many respects analogous to the British. For example, both chose to develop the gas-graphite reactor. The French decision to remain with the gas-graphite reactor would later pit Gallic technology and patriotism against economic reality. The fact that France had no alternative reactor designs only served to intensify the future struggle. The French situation differed in one important aspect from that in Britain. The nascent emergence of the nationalized power utility meant that CEA's authority would not go unchecked.

---

[91] Regarding the latter, see "Progress Report—EL4," *Nuclear Engineering* 11, 121 (June 1966):473.

Possible Ramifications

The national nuclear R&D strategies were linear continuations of the prior conditions and organizations. The United States and the Soviet Union both entered the nuclear power era with large and varied programs that were reflected in their multiple reactors research strategies. Canada and France both operated their nuclear R&D on a much more restricted scale and with a much reduced research base, so it could be expected that they would adopt a single reactor R&D approach. The case of the British and the West Germans is less clear. As Burn points out, the British monetary and personnel expenditures were on the same order of magnitude as those in the United States during the initial development of nuclear reactors.[92] England's single alternative strategy was probably dictated more by the absence of enriched uranium and the motivation to join the nuclear power family as expeditiously as possible than by Britain's manpower, financial, or scientific base. Germany might well have chosen to pursue a single reactor strategy, but the lingering uncertainty present throughout the late 1950s and the intense competition among German private industries argued convincingly to the contrary. In short, the British, French, and Canadian nuclear R&D programs were severely restricted by prior circumstances, of which the most important was the lack of unenriched uranium. The American, Soviet, and the West Germans each had multiple research avenues open to them; rather than unduly restrict themselves in the face of serious technological uncertainties, each nation chose to maintain multiple options until their reactor prototypes or development models provided sufficient information about technical and economic uncertainties so that a choice could be made.

Among the single reactor technology nations, two decided on basically the same reactor, the gas-graphite, and the third chose the heavy water reactor. An interesting difference between these R&D programs was the pace at which the countries developed

---

[92] Burn, *The Political Economy of Nuclear Energy*, pp. 82-84. Although the exact numbers are at contention, the great difference that one might have expected between the two nation's expenditures did not seem to exist.

their reactors. The British prototype commercial power reactor at Calder Hall became operational in 1956 or about ten years after England decided to develop a nuclear facility; the French prototype reactors, Marcoule 1 and 2, became critical in 1959 and 1960, also about a decade after the French decision. The Canadian NPD did not achieve criticality until 1962, about sixteen years after Canada's nuclear decision, even though Canada had been active in atomic research since the wartime agreements and was not distracted by weapons research.

It is accurate to say that these small, initial steps were to bind each of the nations almost inextricably into a certain set of programs. For the United States, Soviet Union, and West Germany, the consequences were not particularly onerous. However, for the French and British, their early commitments were to prove shortsighted, a judgment made even more difficult to accept because of the early hope and promise of their reactor systems. Canada's nuclear reactor program was similarly set but the results were not nearly as unfortunate.

The ramification of the initial research commitments first became obvious as each nation began to disseminate and commercialize its reactors and as the projected costs were held up to reality. The demonstration period is the final period before acceptance or rejection (and attendant redesign) of the nuclear reactor system. It is to this stage, then, that we next move.

6

Reactor Demonstration: Participants and Strategies

Definition and Organization

In many ways, the reactor demonstration period was the period of the greatest similarity or comparability among the six national nuclear reactor programs. Each nation tested a large prototype of the nuclear reactor it thought would best serve its own power purposes given the various technical and resource constraints. These reactors were tested as demonstrations of their operational capabilities and financial viability. But this apparent sameness should not obscure real and important differences among the surveyed nations. Some chose to demonstrate the capabilities of more than one reactor, while others chose to continue their reliance upon a single type. Most built multiple reactor demonstrations, while others were content to rely on a few sites. And each nation displayed an important shift among and within the relevant actors that composed the reactor RD&D coalitions.

In a way, this is the least distinct of the three enumerated periods in reactor RD&D and dissemination. Nations such as Canada and the United States were able to make fairly smooth transitions from the development to the demonstration stage and then to the final stage of successful dissemination or commercialization. However, this stage was illustrative of how national programs demonstrated the products of their technological research and development establishments with a final set of design alterations and evidence before a decision was made to initiate full-scale "commercial" dissemination. Furthermore, a changing set of research priorities and actors was manifested, which could be technological, economic, or political. All of these factors decisively affected the final decision about the widespread application of nuclear reactor technology.

By the time a reactor has reached the demonstration period, its technology is basically settled.[1] Changes are usually only minor modifications or of such a magnitude as to return the program back to its initial prototype stages. Therefore, the emphasis in this chapter is on the altering roles and coalitions of the relevant actors.

Generally, this period was a transition in reactor RD&D from the research community to the user community and from applied research to actual application. Often, private industry and electric utilities facilitated and bridged this transition. However, this process was not always direct. In the Soviet Union, the institutional politics were so obscured as to hide the transition. In France and Great Britain, the role of private industry was still minimal. The movement from a prototype reactor to an acceptable commercial reactor was strained, partially because of this exclusion. Similarly, the role of government research agencies and the electric utilities varied in both degree of involvement and importance.

As in Chapter 5, in this chapter each national experience is examined and comparisons drawn between the national programs at the end. Again, a distinction is made between nations that

[1] Perry et al., *The Light Water Reactor*, p. 86, note that there has not been a single significant technical change in the LWR since the early 1960s. There have been, of course, a number of safety mechanism changes.

demonstrated a single reactor nuclear type (England, France, and Canada) and nations that demonstrated multiple reactors (the United States, Soviet Union, and West Germany).[2]

Multiple Reactor Demonstrations

United States[3]

The most multifaceted nuclear reactor demonstration program was in the United States. As noted earlier, in 1954 AEC decided to emphasize five different reactor types: the PWR, the BWR, the fluid fuel reactors, the sodium cooled reactor, and the breeder reactor.[4] In addition, AEC continued its research on an experimental gas cooled reactor while private industry and the utilities conducted research on other reactor types, such as a heavy water cooled and moderated reactor and a high temperature gas reactor.[5] Thus, this period in the United States was not so much the demonstration of a single reactor technology as it was picking and choosing among an embarrassment of nuclear riches. AEC's role during this period was critical, as it began its transition from the role of nuclear development to the dual roles of regulator and promoter of atomic energy.[6] However, "decentralization of operational responsibilities and readjustment of proprietary relationships do not mean decentralization of governmental regulation and control."[7] That AEC was no longer the sole or primary nuclear development agent did not mean that

---

[2] A schematic chronology of the different national reactor development programs is found in Appendix C.

[3] The American section of this analysis will be cursory because of its exhaustive detail elsewhere.

[4] See Appendix B for an elementary description of reactor design components and operation.

[5] For a brief account of the U.S. development of some of the reactor types that were not later commercialized, see Edward Creutz, "Nuclear Power: Rise of an Industry," *Bulletin of the Atomic Scientists* 25, 6 (June 1970):75-82.

[6] The transition is documented in Elizabeth Rolph, *Regulation of Nuclear Power: The Case of the Light Water Reactor* (Santa Monica, Calif.: The Rand Corporation, R-2104-NSF, June 1977), esp. Secs. III and IV.

[7] Scheinman, "Security and a Transnational System," p. 282.

the government was relinquishing its control over the development of nuclear energy.

Thus, the critical organizational transitions in the United States during this period were the changing roles of AEC from developer to regulator and promoter, of private industry from contractor to entrepreneur, and of the electrical utilities from observer to participant. All these interacted closely with each other and the different reactor technologies or systems to form different coalitions for the delivery of efficient reactor technologies.

In 1953, AEC decided that nuclear power had been sufficiently developed so that the construction of demonstration reactors should be encouraged. In early 1953, Gordon E. Dean, then chairman of AEC, urged Congress "that the development of nuclear power for central-station use be recognized as an important national objective and that the [Atomic Energy Act] law be changed to allow private initiative."[8] This philosophy was affirmed in December when President Eisenhower delivered his "Atoms for Peace" address to the United Nations. That these statements were more than political rhetoric was confirmed when the AEC approved the construction of the 60 MW(e) Shippingport PWR reactor to be built by Westinghouse and operated by the Duquesne Light Company, and Congress passed the Atomic Energy Act of 1954, which made nuclear fuel available to private industry for the first time. The legislation stipulated that the AEC would retain title to the fuel. Allen has commented: "Implicit in the provisions of the 1954 law was the expectation that private industry would begin to invest in nuclear R&D power plant construction, thereby reducing the role the federal government might have had to play."[9] In addition, the "AEC was charged with the regulatory responsibility and given sole authority to grant licenses for the construction and operation of nuclear power facilities and for the possession of nuclear fuel."[10] Finally, the law prohibited AEC from financing or constructing

---

[8] Hogerton, "The Arrival of Nuclear Power," p. 23.
[9] Allen, *Nuclear Reactors for Generating Electricity*, p. 37.
[10] Hogerton, "The Arrival of Nuclear Power."

power reactors whose primary purpose was the distribution of electricity. A JCAE report summarized the intent of the new act:

> Many technological problems remain to be solved before widespread atomic power, at competitive prices, is a reality. It is clear to us that continued Government Research and Development using Government funds will be indispensable to a speedy and resolute attack on the problems. It is equally clear to us, however, that the goal of atomic power at competitive prices will be reached more quickly if private enterprise, using private funds, is now encouraged to play a far larger role in the development of atomic power than is permitted under existing legislation. In particular, we do not believe that any developmental program carried out solely under governmental auspices, no matter how efficient it may be, can substitute for the cost-cutting and other incentives of free and competitive enterprise.[11]

However, private industry was generally unwilling to apply for nuclear construction permits, and therefore in January 1955 AEC announced the Power Reactor Demonstration Program (PRDP) to encourage private industry to become a participant in nuclear RD&D and to stimulate the demand for nuclear reactors. Under PRDP, AEC offered to waive the normal fuel inventory use charges for the first five years of the reactor's operation and to provide extensive R&D and technical assistance. In return, the utilities would have to finance, build, and operate the plants themselves. PRDP resulted in four construction applications: Yankee Nuclear Power Station, a PWR built in Rowe, Massachusetts; an experimental liquid sodium cooled reactor at Hallam, Nebraska; an application from Commonwealth Edison that was later withdrawn; and the Enrico Fermi Atomic Power Plant at Lagoona Beach, Michigan, the first "fast" reactor built for the commercial production of electricity. Three other nuclear facilities were announced that were financed with private funding: the Vallecitos BWR at Pleasanton, California, which would be built by General Electric as a test facility; the Dresden BWR Nuclear Power Station at Morris, Illinois, which would be

---

[11] U.S. Senate, 83d Cong., 2d Sess., "Amending the Atomic Energy Act of 1946 as Amended and for Other Purposes," in *Legislative History of the Atomic Energy Act of 1954* (P.L. 83-703), Senate Report No. 83-1699, Washington, D.C., 1951, vol. 1, p. 751; cited in Allen, *Nuclear Reactors for Generating Electricity*, p. 38.

operated by Commonwealth Edison; and the first Indian Point PWR reactor, at Indian Point, New York, which was to be built for Consolidated Edison. With the exception of the Yankee reactor, the PRDP units were test reactors and were designed to generate only very small amounts of electricity, if any; the Vallecitos BWR was similarly experimental. Yankee, Dresden, and Indian Point were designed to generate 140 MW(e), 180 MW(e), and 151 MW(e) (which was increased to 265 with fossil-fueled superheating of the steam), respectively.

By late 1955, AEC was still not satisfied with the electric industry's modest acceptance of nuclear reactors. In particular, AEC was confronted by a recalcitrant utility industry and armed with little leverage of its own. Scheinman noted this as one critically important characteristic of the American situation:

> Where power production is not nationalized, the capacity of the government to exercise promotional responsibility and to influence the course of events is largely confined to setting national objectives and to creating conditions and providing incentives likely to foster these objectives; it cannot impose the use of nuclear power to the exclusion of alternative energy sources.[12]

Even so, the government and the AEC had a persuasive—if not coercive—arsenal of incentives available,[13] not the least of which was the implied threat to build and operate its own nuclear power facilities. The TVA experience had not been forgotten by the utilities. Hogerton summarized their dilemma at the time:

> Since it was clear that the Government was committed to a policy of [nuclear reactor] acceleration, the investor-owned segment of the electric-utility industry, representing about 90 percent of the total electric industry, faced a difficult choice—to build or not to build. To build full-scale noneconomic nuclear plants promised to be a costly way of speeding the development of a technology for which only a few farsighted executives could see a pressing need. On the other hand, not to build meant running the risk of having the Government do the job itself. In a collective sense what the utility companies did at that critical time was to follow a middle course: they committed themselves to building a sufficient number of plants and took enough other in-

12 Scheinman, "Security and a Transnational System," pp. 281-282.

13 The AEC's policy leavers are reviewed by Norman Hilberry, "Nuclear Power in the U.S.," *Nuclear Industry* 11, 8 (August 1964):5-12.

itiatives to maintain their position with respect to the Government in an accelerated effort.[14]

Even though the first PRDP program was less than a year old, AEC announced a second in September 1955. The so-called "second round" PRDP was "designed to encourage the construction of small-scale prototypes of novel reactor systems and also to facilitate the participation of small publicly owned power entities such as rural cooperatives and municipal power authorities."[15] Under the second PRDP, AEC offered to finance the reactor part of any nuclear reactor, thereby limiting the financial risk of a participating utility, as well as continuing the incentives offered for the first program in which fuel costs were waived and technical aid was provided. An additional five proposals responded to these inducements.

By 1957, the first results of AEC's demonstration strategy began to be evident. In August, the Vallecitos BWR facility, the experimental prototype reactor for GE's Dresden nuclear reactor, went critical. Vallecitos allowed the GE engineers to test the BWR's operational characteristics and to gather data pertinent to the larger, commercial Dresden reactor design, which was under construction. In December, Shippingport went into operation and soon achieved its full power rating of 60 ME(e). In 1960, both Dresden and Yankee went critical and operated according to expectations. Nevertheless, the electric utilities were reluctant to commit their resources to nuclear power. Their rationale was quite reasonable. Even though demonstrations of nuclear power reactors were increasingly providing solid indication of nuclear reactors being technically feasible and operationally viable, the economics of nuclear energy versus fossil fuels was uncertain at best and still unfavorable by most accounts. With the exception of the Yankee power station,[16] most of the early reactors were

---

[14] Hogerton, "The Arrival of Nuclear Power," pp. 23-24.

[15] Ibid., pp. 23-24; see Allen, *Nuclear Reactors for Generating Electricity*, Sec. IV, for an account of the forces that led to the second round PRDP.

[16] Yankee was built on schedule and at less than the contract price, a unique phenomenon in reactor construction. See William R. Hederman, Jr., "Yankee Nuclear Power Reactor and Connecticut Yankee Nuclear Power Reactor," in Baer et al., *Analysis of Federally Funded Demonstration Projects: Supporting Case Studies*, pp. I-1 to I-32.

costing more to build than forecast, and the operational costs were close to 50 percent greater than initially estimated. In addition, coal costs were declining as the coal industry prepared to meet the challenge of the new power source.

AEC again attempted to stimulate the nuclear reactor market. In 1957, Congress passed the Price-Anderson Act amendment to the 1954 Atomic Energy Act. Before this legislation, no single insurance company or even an insurance pool could afford to insure a utility against liability in case of a nuclear accident. The new act required the licensee to provide only $60 million worth of insurance coverage for public liability claims; the government assumed responsibility for the rest up to a maximum amount of $560 million set by AEC.[17] "Guaranteed against financial loss in the event of a nuclear accident, private utilities began to move into the nuclear field with only the question of economic competitiveness of nuclear power standing between them and nuclear reactors."[18]

In 1960, as a result of AEC inducements and increasing private industry involvement, construction was begun on a number of power-producing, prototype, and technology demonstration plants. These included an experimental Westinghouse PWR in Saxton, Pennsylvania; a 48.5 MW(e) BWR in Eureka, California, which was to be operated by Pacific Gas and Electric; another BWR at Big Rock Point, Michigan, rated at 47.8 MW(e); the Carolinas-Virginia Tube Reactor (CVTR), in South Carolina, a 17 MW(e) heavy water cooled and moderated reactor; and an experimental organic liquid cooled reactor at the National Reactor Testing Site, Idaho.[19] In addition, a HTGR, rated at 40 MW(e), was under construction by Gulf-General Atomic at Peach Bottom, Pennsylvania.[20] By the end of 1962, there were eleven government-owned reactors, twelve reactors jointly owned by

---

[17] The issues of nuclear insurance are addressed in Sheldon Novick, *The Careless Atom* (Boston: Houghton Mifflin, 1969); also see Rolph, *The Regulation of Nuclear Power*, Sec. III.

[18] Scheinman, "Security and a Transnational System," p. 283.

[19] See U.S. Atomic Energy Commission, *Report to Congress, 1960*, January 1961, Washington, D.C., p. 18.

[20] HTGR reactor research by Gulf-General Atomic is described by Creutz, "Nuclear Power: Rise of an Industry," p. 78.

AEC and a utility, and seven reactors initiated by private industry operating in the United States. Each was considered to be an experimental or demonstration facility.[21] In addition, significant amounts of operational data on large reactors were being collected as Dresden, Yankee (which became critical in 1962), and Shippingport (1957) supplied daily electricity to their customers. Indeed, by the end of 1962, Shippingport had completed five years of operation, supplying 1.3 billion KwHr of electricity to Pittsburgh.[22]

In 1963, four new nuclear stations became critical, including the CVTR and two breeder reactors.[23] This coincided with AEC's 1962 report to the president on the Civilian Nuclear Power Program, which concluded that "nuclear power had arrived at the threshold of commercial application and had a bright future."[24] Its recommended program included the construction of LWR facilities on a commercial scale.[25] This position was reiterated even more forcefully the following year when AEC announced that "the light water, saturated steam reactors are the most highly developed at this time and future Government support will be largely limited to conducting economic and technical analyses of operating reactors."[26] Yet, between 1955-1956 (when Dresden, Indian Point, and Yankee were ordered) and 1963, no electric utility placed an order for a commercial-size nuclear power station with the exception of Connecticut Yankee at Haddam (late 1962).[27] AEC, in conjunction with private industry and the electrical utility companies, had demonstrated the technical and operational feasibility of a number of different reactor types. However, to date no reactor design had really taken hold and generated enough utility support to warrant being considered

---

[21] See U.S. Atomic Energy Commission, *Report to Congress, 1962*, January 1963, Washington, D.C., p. 128.

[22] Ibid., pp. 129-130.

[23] Ibid., pp. 75, 86, 91-92, 103-105.

[24] Hogerton, "The Arrival of Nuclear Power," p. 25.

[25] See U.S. AEC, *Report to Congress, 1962*, p. 28.

[26] U.S. Atomic Energy Commission, *Report to Congress, 1963*, January 1964, Washington, D.C., p. 79.

[27] See U.S. Atomic Energy Commission, "Central Station Nuclear Plants," *The Nuclear Industry, 1974*, WASH 1174-73, Washington, D.C., September 30, 1974, Table 1.1, p. 8.

commercially acceptable. The question in 1963 was whether the nuclear power industry would "take off" on its own or whether it could survive only with heavy government subsidization.

During the demonstration period, AEC's reactor research programs and PRDP obviously supported nuclear reactor RD&D. At the same time, the emergence of other powerful actors cannot be ignored. Private industry became increasingly important. Westinghouse, based upon its PWR research for submarine propulsion and the Shippingport reactor, was a key participant; GE built its own BWR test facility at Vallecitos as well as the Dresden plant; Gulf-General Atomic actively pursued a HTGR; and the traditional utility heavy construction firms, such as Babcock and Wilcox and Combustion Engineering, began to enter the industry and to define their roles (e.g., Babcock and Wilcox was the contractor for the first Indian Point reactor facility). The electric utilities also started to assert themselves in both positive (i.e., several prototype reactor plants were ordered by utilities) and negative modes (i.e., few were willing to act as industry leaders and order full-size power reactors). The prototype and demonstration reactors were to provide the operational and cost basis for future decisions on reactors.

Hence, we can see that the relative leadership position of AEC declined as other organizations and institutions became involved. Although AEC clearly had a great deal of influence, the ability of the utilities to resist government encouragement suggests that AEC was no longer the predominant nuclear institution. Indeed, AEC was not able to maintain its control over government participation. In July 1957, Congress amended the 1954 Atomic Energy Act; AEC was now required to receive specific congressional approval of funds for any nonmilitary reactor that was to produce electricity (or had a thermal rating greater than 10 MWs) and for any PRDP project. JCAE was mandated to define the criteria for fuel use charge waivers. In short, AEC was largely relieved of its role as primary custodian of nuclear energy RD&D as it became the enthusiastic booster and regulator of the emerging nuclear industry.[28] Although it is beyond the scope of

---

[28] See Rolph, *Regulation of Nuclear Power*, Secs. III and IV.

this book to elaborate, these new and dual roles of AEC were later to prove serious organizational weaknesses.[29]

Soviet Union

The Soviet Union appears to have moved from its initial prototype reactors into the demonstration phase with only minor impediments. With the exception of the Melekess BWR, the Soviets do not appear to have expended research efforts on any other prototype slow fission reactors than the *Lenin* and Novovoronezh PWRs and the Beloyarsk LWGR, which was an extension of the Obninsk reactor technology[30] (see Table 5-1).

This seeming ease of progression is more apparent than real. Although the factual record is clouded, there is some evidence that the Soviet progression from prototype developmental reactor to the full-scale demonstration of nuclear power was nowhere near as direct as might first appear. For example, as part of the Sixth Five-Year Plan revealed in 1956, Soviet planners announced a total capacity of 2,000 to 2,500 MW(e) to be generated by nuclear power by 1960. The confusion and contradictions among the various Soviet officials made clear that the expected totals were not realistic. In 1957, the Sixth Five-Year Plan was officially abandoned and most of its industrial goals were extended as part of a new Seven-Year Plan that was scheduled to begin in 1959. Specific new nuclear power goals were not released. One of the reasons for this apparent retreat was probably the realization that atomic power was not proving to be economically competitive with alternative power sources. Deputy Premier Frol Kozlov suggested as much during a trip to the United States in 1959.[31] The economics of nuclear energy were not quickly solved

---

[29] For example, Bupp and Derian argue that because of its early organizational emphases, AEC was inherently unable to assume these new responsibilities efficiently; see Bupp and Derian, "The Nuclear Power Industry," pp. 97-100. This issue is also addressed in Perry et al., *The Light Water Reactor.*

[30] U.S. Atomic Energy Commission, Division of Reactor Development and Technology, *Soviet Power Reactors*, WASH-1175, Washington, D.C., August 1970, p. 36.

[31] See Kramish, *Atomic Energy in the Soviet Union*, pp. 144-145.

by the Soviets. Petrosyants confirmed that the Beloyarsk-1 reactor experience (which became operational in 1964) was not fiscally encouraging: "Because of the relatively small installed power of the unit (100 mw), and complicated and expensive technological channels, the cost of 1 KwHr by this plant is high (somewhat higher than the cost of electric power generated by other thermal power plants of the Urals)."[32] Petrosyants makes similar cost statements about Novovoronezh-1.[33] Wilczynski writes that "For a long time the cost of nuclear power appeared to be prohibitive in comparison with the conventional sources. It was thought that further R&D were essential and that nuclear power could not become competitive before 1980."[34]

Finances were not the only obstacles Soviet nuclear planners faced. There is evidence of skepticism within the Soviet heavy construction sector. Petrosyants has commented, "True, there was a time when this [nuclear] technology was regarded with misgivings. Production people were reluctant to try it, and we atomic researchers had to make the equipment ourselves and present it free of charge to plants so that they might see for themselves that it was reliable and safe."[35] Even though the Soviet nuclear administrators and technicians have not been particularly open about the technical failures of their programs, it is clear that there were some serious shortcomings that required correction. Petrosyants admitted that "the reader should know that during the start-up and initial operation of this [Novovoronezh-1] power plant not everything ran quite smoothly. Unavoidable troubles and excitement [!] took place during the period of the start-up and adjustment operations caused by

---

32 Petrosyants, *From Scientific Search to Atomic Industry*, pp. 92-93. Petrosyants' continued attention to reactor costs should dispel any thought that the Soviet R&D program was not as responsive to economic constraints as those in capitalist societies.

33 Ibid., p. 114.

34 Wilcznski, "Atomic Energy for Peaceful Purpose . . . ," p. 570. Also see Polach, "Nuclear Power in East Europe," p. 11.

35 Andranik Petrosyants, "The Atom on the Scales of Progress," *New Times* 13 (March 1976):18-19. Petrosyants was responding to an interview conducted by Helen Knorre. The "free of charge" expression is particularly provocative.

phenomena which were not immediately explained."[36] Even more serious, there are partially substantiated reports that a nuclear "explosion" took place in the Urals in 1958 and released large amounts of radioactive materials into the atmosphere. Soviet dissident scientist Zhores Medvedev claims:

> Strong winds blew the radioactive clouds hundreds of miles away. . . . Tens of thousands of people were affected, hundreds dying, though the real figures have never been made public. The large area, where the accident happened, is still considered dangerous and is closed to the public.[37]

Soviet spokesmen have never acknowledged Medvedev's charge. He is supported by American intelligence sources, which reportedly picked up traces of the radioactivity released by the explosion.[38] The causes of the explosion are not known. Medvedev writes that it "was an enormous explosion, like a violent volcano," created by nuclear wastes buried in shallow disposal pits.[39] Others suggest that the accident involved a plutonium-producing reactor that went out of control and ex-

---

[36] Petrosyants, *From Scientific Search to Atomic Industry*, p. 113. Petrosyants provides examples of a few Soviet technical difficulties.

[37] Zhores Medvedev, "The Two Decades of Dissidence," *New Scientist* 72, 1025 (November 4, 1976):265. Medvedev is a noted critic of Soviet scientific policies; at the time the article was written, he was also active in opposing a British shallow-burial nuclear waste disposal plan. Further evidence is provided by Andrew Cockburn, "The Nuclear Disaster They Didn't Want to Tell You About," *Esquire* 89, 7 (April 25, 1978):39-43.

[38] Robert Gillette, "Russ Atom Accident Laid to Reactor," *Los Angeles Times* (November 10, 1976), pp. 1, 10, cites "American intelligence experts" who support Medvedev's claim of released radioactivity but question his casualty estimates. "As for why the U.S. Government had kept knowledge of the Soviet accident secret, several officials said it was to avoid indicating the sensitivity of the U.S. atomic test detection system. And one source noted that 'There was apparently no inclination to rub their noses in it.' " Robert G. Kaiser, *Russia: The People and the Power* (New York: Athaeneum, 1976), p. 313, also reports the explosion.

[39] Medvedev, "Two Decades of Dissidence." Sir John Hill, chairman of the U.K. AEA, dismissed Medvedev's charges as "rubbish," claiming that the type of waste buried had a low activity rate and "could not possibly give that sort of explosion." He also said that he had never heard any suggestions of such an explosion from his contacts in the Soviet Union. Hill is quoted in "Soviet Dissident Discloses Holocausts," *Aviation Week & Space Technology* 105, 20 (November 15, 1976):20. Cockburn, "The Nuclear Disaster . . . ," after sifting the various charges, strongly supports Medvedev.

ploded.[40] In either case, but particularly the latter, the accident probably slowed down the Soviet civil reactor program, although it seemed to have little effect on the pace of their weapons program.[41]

The first electricity-producing LWGR was the Obninsk plant, but it was too small for any operational data to be obtained. To gain those data, the Soviet nuclear engineers built the 100 MW(e) Beloyarsk-1 reactor, which became critical in 1964. Although the technology was similar to the Obninsk reactor, the Soviets apparently intended Beloyarsk-1 as a demonstration-sized reactor to give its technicians operational experience as well as to provide information about nuclear technology and costs. Important technical advances were included. For example, Beloyarsk-1 was not built with a heavy, multiton reactor pressure vessel that an equivalent PWR would have had. Petrosyants implies that this advance was driven by construction economics: "The possibility of constructing nuclear power plants with channel-type reactors without a reactor pressure vessel is rather attractive, since the heavy industry is released from manufacturing steel pressure vessels weighing 200 to 250 t[ons]."[42] Again, the costs of Beloyarsk-1 generated electricity were not favorable compared with other sources. The technical difficulties and financial disadvantages of this reactor might have inhibited the Soviet nuclear industry except that Beloyarsk-1 was rather clearly designed and operated as a test reactor. This assumption is supported by the opening of the second Beloyarsk reactor approximately three and a half years after the opening of Beloyarsk-1. The short time lag between the two reactors strongly suggests that Soviet nuclear policymakers had decided during the construction of the first reactor that a second, larger reactor would be required for ac-

---

[40] Gillette, "Russ Atom Accident Laid to Reactor," p. 1.

[41] One of Gillette's sources is quoted as saying, "There is no evidence that the accident put a crimp in their plutonium production or that it gave them second thoughts about their weapons programs. They just piled dirt over the reactor—buried it that way—and went about their business." (p. 10) Kaiser, *Russia: The People and the Power*, seconds this assessment: "Early experiments resulted in tragic accidents . . . causing hundreds of deaths, but nothing slowed the Soviet push for atomic weapons," (p. 313).

[42] Petrosyants, *From Scientific Search to Atomic Industry*, pp. 89 and 92.

curate evaluation of the LWGR's potential for nuclear-generated electricity. Knorre summarizes the Soviet reactor demonstration philosophy:

> Although reactors of this type do have a number of shortcomings their continued construction permits accumulation of experience in the building, mounting and manufacture of reactor equipment. They also help to solve a number of problems connected with the creation of large industrial atomic power plants.[43]

Thus, one can assume that the Beloyarsk reactors were not being evaluated on the basis of strict cost or efficiency; in other words, they were technology demonstration facilities.

Beloyarsk-2, rated at 200 MW(e), featured a somewhat simpler, more reliable design.[44] It operated at a marginally more efficient rate; the thermal efficiency of the second plant was 37.8 percent rather than 36.3 percent for the first. The operations and cost information garnered from these reactors seemed to prove to the Soviets that "channel-type reactors with a graphite moderator, cooled by ordinary water, should be considered as reliable and economically profitable for utilization at nuclear power plants, especially at the powerful ones."[45]

The Soviet demonstration PWRs were at Voronezh on the River Don. At the end of September 1964, Novovoronezh-1, rated at 210 MK(e), was declared in operation; by December, it had reached its rated capacity. Besides demonstrating a different reactor technology, it also was designed to incorporate new materials technology, such as zirconium alloys for the fuel element cans.[46] Even though "unavoidable troubles and excitement took place" during the initial operations of Novovoronezh-1 and the costs for PWR generated electricity were higher than for coal, the Soviets were encouraged by its performance. This evaluation was strengthened as the costs per kilowatt hour steadily declined: the Novovoronezh-1 cost per KwHr was equal to 1.24

---

[43] Knorre, "Frontiers of Technology: U.S.S.R.," p. 317.

[44] A diagrammatic comparison of Beloyarsk-1 and -2 is provided in Petrosyants, *From Scientific Search to Atomic Industry*, pp. 92-93.

[45] Ibid., p. 96.

[46] Soviet work on zirconium alloys is particularly noted in "W.H. Zinn Evaluates Russian Reactor Progress," p. 104.

kopecks in 1965, 1.22 in 1966, 1.18 in 1967, 0.96 in 1968, and 0.90 in 1969. (This compared with 0.8 kopeck per KwHr for a neighboring coal-powered station, or as low as 0.5 to 0.6 kopeck per KwHr at some of the more modern Soviet coal-burning stations.)[47] Again, as with Beloyarsk-1, the Soviets were not deterred by the costs and operational difficulties encountered at Novovoronezh-1. Petrosyants explains that "it should be borne in mind that the first power unit of the Novovoronezh NNP, and the whole power plant, was constructed not only as a commercial complex, but as a demonstration facility as well."[48]

One can speculate that the Soviets were less convinced about the viability of the PWR than the LWGR, perhaps because the LWGR had an operational history dating back to 1954. The briefer experience with the PWR also meant that there was less time to solve the various problems that occurred. These two factors might explain why Novovoronezh-2 did not reach its operational status until December 29, 1969, five years after the first Novovoronezh reactor. The time lag between Beloyarsk-1 and -2 was a relatively brief three and a half years. Novovoronezh-2 had an electrical capacity of 365 MW(e), but the reactor core and pressure vessel were practically the same as with the first unit, which implies the Soviet nuclear design teams learned a great deal from the first Novovoronezh PWR. In January 1969, improvements on the first unit were initiated that raised its MW(e) capacity to 280, an increase of 34 percent. In combination, the two plants were generating electricity at a cost of 0.8 kopeck per KwHr in 1971.

The demonstration of the PWR and LWGR technologies, combined with the development but not demonstration of the BWR, implies two important characteristics of the Soviets' R&D strategy during the demonstration period. First, Russian nuclear planners appeared to be unwilling to commit the resources necessary to continue the development of many reactor types. Rather, an early winnowing among reactors reduced the Soviet

---

47 The cost figures are cited in Petrosyants, *From Scientific Search to Atomic Industry*, p. 115. There is no indication if the kopecks are adjusted to a given year's value.

48 Ibid., p. 115.

nuclear R&D to the development of three slow fission choices, with the PWR and LWGR receiving the greatest emphasis. This decision was made on the basis of experimental calculations and some prototype reactor data. Heavy industry might have affected this decision. Second, and more important, once the Soviets had made this choice to build demonstration PWR and LWGR reactors, it was obvious that they had made a major commitment to these nuclear reactors. Soviet decisionmakers gathered a great deal of operational and cost data for both types of reactors. At least two reactors of each type, one of which could be considered a demonstration of a full-sized commercial-sized reactor, were planned and built. In addition, the Soviets apparently chose to collect time series data on these reactors. Limited data or information collected from demonstration reactors was not considered sufficient for evaluating the reactors at this stage of their development. Petrosyants explained this conservative approach:

> The development of the nuclear power industry cannot be based upon the designing experience only, even confirmed by experimental data and testing. Moreover, even the operation of separate large reactor prototypes is not a sufficient justification for a wide utilization of a given reactor type. Only the generalized multi-years' experience gained by industry as a whole can serve as a reliable foundation for the large-scale nuclear power industry.[49]

By the late 1960s the Soviets seemed to have obtained sufficient data on the PWR, LWGR, and perhaps the BWR to permit them to make a decision about the future dissemination of nuclear power reactors, both for domestic use and as an export to the nations of the Eastern European bloc. The overall picture of the Soviet reactor program showed steady, deliberate, and focused growth with a gradually increasing specific power of the core design. The two main areas of concern seem to be reliability and safety.[50]

---

49 Ibid., pp. 97-98.

50 Ibid. These two primary areas of concern in Soviet reactor development are also stressed by Pryde and Pryde, "Soviet Nuclear Power"; and Gillette, "Nuclear Power in the U.S.S.R." Petrosyants, "The Atom on the Scales of Progress," pp. 19-20, offers a lengthy description of Soviet nuclear reactor safety concerns and measures.

Unfortunately, the lack of information about the Soviet Union's bureaucratic infrastructure makes any description of the changing roles of the various Russian atomic energy organizations highly speculative.

Federal Republic of Germany

The West German example presents a remarkable institutional contrast to the American demonstration period. During this period, the role of the government in West Germany was increased, largely due to the demands of German industry. As noted earlier, the German Ministry for Atomic Energy (MAE) was formed in 1956 to coordinate regional research on atomic energy, and the German Atomic Commission (DAK) was little more than an advisory committee. The central government adopted a low profile regarding nuclear energy. With the exception of partially funding the research centers at Jülich and Karlsruhe and issuing the unofficial Five-Year Eltville plan, the central government played only a supportive role in the growth of the German nuclear industry through the prototype reactor stage, which lasted until the early 1960s in Germany. In the early 1960s, the federal government increased its research funding and began to offer larger amounts of financial support to the Eltville program, which had emphasized German research on the LWR, the HTGR, and the heavy water cooled reactor. Still, by the mid-1960s, it was clear to many in Germany that West Germany was falling behind in nuclear technology and that private industry could not carry the entire load.

At the same time, it became obvious that the German utilities were not enthusiastic about nuclear power for several reasons. First, Germany had over 350 electric utility companies, whereas most of its European neighbors had one or two nationalized electric utilities. Only 3 percent of the German utilities were predominantly owned by private enterprise; 42 percent were financed by a combination of federal, state, and local governments, and the remaining 55 percent were public-private mixes in which public authorities held the majority interest. This division could have restricted the utilities' access to the capital necessary to finance nuclear reactors. However, the market was not as fragmented as first appeared. One utility, RWE, produced close to 30 percent of Germany's total electricity, the twelve largest utilities produced 65 percent, and thirty-one companies generated 90 percent of the total supply of electricity.[51] Second, many utilities were not eager to adopt nuclear power because "the largest German utilities operated their own coal mines and faced internal obstacles in developing a direct competitor to coal-fueled power plants."[52] Finally, under the federal German energy law, utilities were compelled "to provide power with maximum availability at minimum costs."[53] This requirement was cited to deflect government insistence that the utilities enter into the long-term capital commitments required for nuclear generating plants. Hence, of the German utilities, only RWE and ABK were strongly disposed toward nuclear-generated electricity as this period began in the early 1960s.

The dual forces of pressure from private industry and the reluctance of a fragmented, nonnationalized utility industry meant that the German government had to assume an active participation in the development of nuclear energy if West Germany were to compete in this area. MAE and DAK, however hesitantly, increased the central government's involvement in the demonstration of nuclear energy.

The capital costs of nuclear power were such that nuclear reactors were extremely difficult to build without government assistance. Given the minimum participation of the central government, West Germany did not have a demonstration size nuclear reactor under construction by 1961. This was in stark contrast to the British, French, Italian, and even the Belgian nuclear construction programs. In addition, the paucity of government funding had impeded domestic reactor research. Therefore, when the decision was made that a nuclear plant be built at Gundremmingen, it was not surprising that it was not a *Bund*-financed program; rather, it was a joint product of two

---

51 W. Mackenthun, "Survey of Germany: Electric Supply Industry," *Nuclear Engineering* 13, 148 (September 1968):745-748.

52 Nau, *National Politics and International Technology*, p. 87.

53 "Germany," December 1960, p. 543.

utilities, RWE and AKB. Their choice was between two foreign reactor types, a British gas-graphite reactor and an American BWR reactor, each to be built by German firms under foreign licenses. The contract ultimately went to GE/AEG to construct a 250 MW(e) BWR plant to be fueled with American enriched uranium.

This decision was critical for the German nuclear industry. RWE executives favored the American reactor design chiefly because of the lower initial investment capital required for the LWR.[54] The BWR proposed by GE for the Gundremmingen plant called for a capital expenditure of $88 million, compared with $133 million for the English Electric gas-graphite reactors.[55] Just as important was the American willingness under the EURATOM joint program to guarantee enriched uranium, which was to bias the German nuclear industry toward the American design.[56] Nau elaborated upon the significance of this:

> The existence of this program altered expectations in favor of American reactors from 1958 on. Europeans were assured of a hand in research results in the United States and U.S. guarantees of enriched uranium supplies removed a primary obstacle to light water developments in Europe. Without these provisions, European authorities would possibly have had to invest more attention and resources in gas graphite reactors. And with enough investment, these reactors may have reached an economical stage by the time of the Gundremmingen bidding equivalent to that of the light water systems.[57]

The Gundremmingen decision was doubly significant because it caused one of Germany's largest construction firms, Siemens, to alter its nuclear development priorities from natural to enriched reactors.[58] At the time, France was laboring strenuosly to

---

[54] See Finke, "F.R.G. Research Activities," p. 743.

[55] See *Nucleonics Week* (May 24, 1962), p. 5.

[56] Some have referred to EURATOM as "A Trojan Horse for Light Water;" Irwin C. Bupp and Jean-Claude Derian, *Light Water: How the Nuclear Dream Dissolved* (New York: Basic Books, 1978), ch. 1.

[57] Nau, *National Politics and International Technology*, p. 145; also see Finke, "F.R.G. Research Activities," p. 743.

[58] The decision by Siemens was not lightly taken. Its research efforts were already enjoying some success in the international market; Siemens was to beat out five other nations when Argentina chose its heavy water PWR design in February 1968.

wean Europe, especially Germany, away from the United States. Largely because of the RWE decision to adopt the LWR, Germany (which was represented by Siemens) and France were not able to agree on a proposed Franco-German 500 MW(e) gas-graphite reactor to be built at Fessenheim. Siemens' decision against the gas-graphite reactors resulted in a serious strain on Franco-German relations in the early 1960s. This was the opening volley of what came to be known as *la guerre des filières*, the reactor war, in Western Europe.[59]

The central government still chose to support German industry's efforts to develop nuclear plants, rather than to lead the way. It actively encouraged their cooperation with American industry as a means of demonstrating nuclear power in the Federal Republic at the lowest cost in capital expenditures and development time. In 1962, Siemens renewed and broadened its licensing ties with Westinghouse and AEG did likewise with GE in 1964. Both companies were persuasive advocates. The reliance on U.S. nuclear technology was reaffirmed when the next two German orders for nuclear plants were contracted to AEG for a 268 MW(e) BWR at Lingen and to Siemens for a 345 MW(e) PWR at Obrigheim (see Table 5-2). These orders signaled the acceptance of the LWR by German electric utilities. In 1965, after a series of additional LWR orders, MAE announced that further financing and development on the LWR would primarily be industry's responsibility. "After this point, federal research programs concentrated on longer-term projects including advanced reactors and the development of nuclear fuel facilities."[60]

Private firms also exercised the primary responsibility for fuel fabrication. Whenever possible, the federal government continued its policy of supporting nuclear research through contracts to private industry and through its association with EURATOM and ENEA. The political reasons underlying this posture were very evident during the *Bundestag* debates on the

---

59 American participation in this struggle is described by Warren H. Donnelly, *Commercial Nuclear Power in Europe: The Interaction of American Diplomacy with a New Technology* (Washington, D.C., prepared for the Subcommittee on National Security Policy and Scientific Developments of the Committee on Foreign Relations, U.S. House of Representatives, December 1972).

60 Nau, *National Politics and International Technology*, p. 87.

nonproliferation treaty (NPT) and German fears of potential discrimination against nonnuclear nations.[61] Indeed, Dr. Balke even suggested that "the slow rate of [F.R.G.] progress in nuclear energy has been purposeful in order to allay any suspicions there might be of the country attempting to follow an objective it had renounced."[62]

Germany had important research relationships with the two European advanced reactor projects, EURATOM's ORGEL[63] (an organic-liquid cooled reactor) and the British-ENEA DRAGON[64] (an advanced high temperature gas reactor). ORGEL was originally conceived in 1958 as part of EURATOM's first five-year program. Research with organic liquids as coolants was selected as a default option because Germany (specifically Siemens) was already conducting research on heavy water cooled reactors and the French argued against additional development gas cooled reactors, which might undercut the CEA laboratories. DRAGON was ENEA's advanced high temperature gas reactor, but it was primarily staffed by British nuclear technicians. DRAGON, the common reactor research program in ENEA's second five-year plan, was agreed upon in March 1959 but did not get under way until 1962. Germany had conducted extensive HTGR research at Jülich and was hesitant to share its advanced technology, especially on the pebble-bed fuel concept. However, by 1961 the U.S. development of coated particles and the lack of federal government funds made the German association with DRAGON much more acceptable. To avoid duplication of the DRAGON program, the Jülich HTGR project redirected its efforts to a high temperature, thorium-fueled reactor (THTR), largely because MAE wished to continue its development of the pebble-bed concept.

---

61 For a discussion of the European perspective of the NPT, see Lawrence Scheinman, "Nuclear Safeguards, the Peaceful Atom, and the IAEA," *International Conciliation* 572 (March 1969):38-42; and Donnelly, *Commercial Nuclear Power in Europe*, ch. 11.

62 Quoted in "Germany—A Review," p. 441.

63 From the French, *ORG*ganique *E*au *L*ourde; see J.C. Leny, "ORGEL—A European Concept," *Nuclear Engineering* 6, 67 (December 1961):508-512.

64 DRAGON is a British code word rather than an acronym; its technology, purpose, physical plant, and organization are described at some length in *Nuclear Engineering* 5, 50 (July 1960):299-316.

By the end of 1965, the German MAE strategy of encouraging industrial and utility participation seemed to have been successful. The editors of *Nuclear Engineering* were moved to write:

> Say no more of Germany's late start, for this has been caught up. The comprehensive programme which many thought too broad and ambitious to be accomplished with the time and effort available is being achieved rather than cut back. And Germany has shown faith in the practical future of nuclear energy by placing more power station orders during the past year than any other country.[65]

Review

Thus, in West Germany, the United States, and the Soviet Union, the multiple reactors development strategy adopted earlier was continued into the demonstration phase but with a reduced number of reactor types. The U.S. nuclear industry built demonstration facilities for at least two reactor types, the PWR and BWR (four if one includes the HTGR at Peach Bottom and the Fermi EBR), the Soviets demonstrated large PWR and LWGR reactors (and a small BWR), and the Germans built PWRs, BWRs, a heavy water reactor, and the HTHR.

Three important differences among these national programs should be noted. First, there was a marked transition among the various American participants. AEC surrendered its dominant development position while the large construction companies and the public utilities began to assume significant roles. In West Germany, the central government was forced to play a more active role, but as the NPT debates reflect, the major nuclear actors in West Germany were still the large construction companies, which both fabricated the reactors and performed the primary research tasks.

Second, America and Russia both developed most all of their own nuclear technology indigenously, but the Germans basically imported most of theirs from the United States. This was partly because the Germans had a belated start in reactor development,

---

[65] Editorial, "Germany," *Nuclear Engineering* 10, 112 (September 1965):317. Also see Joachim Pretsch, "Recent Trends in the German Nuclear Program," *Nuclear Engineering* 10, 112 (September 1965):323.

partly because the central government consciously chose to avail itself of its EURATOM participation, and partly because West Germany used foreign nuclear technology and purchases as a bargaining device in the internecine struggle between France and the United States for German allegiance. However, the most likely reason for the German demonstration strategy was to utilize external research as an interim measure while German nuclear facilities developed their own domestic nuclear research and industrial base. This decision was reflected by the construction of LWRs juxtaposed with the continued German research in such newer nuclear technology as the advanced gas reactors and the THTR. In contrast, once the United States and Soviet Union had demonstrated the operational capabilities of their preferred reactors to their satisfaction, the government-sponsored research on alternative slow fission reactors was seriously curtailed. Each nation continued its research on fast breeder reactors.

A third difference was the motivation of the electric industries to accept nuclear power plants. In the Soviet Union, of course, this was a problem in "bureaucratic politics" but susceptible to government ukase. In the United States and West Germany, the fractionated electric utility network was immune from government coercion. In the United States, only extensive persuasion and lucrative government subsidization were available to the AEC as policy levers with which to motivate utility involvement; during this demonstration period, AEC success was hardly conspicuous. In Germany MAE played a less active role, and the principal promoters of nuclear-generated electricity were the large nuclear construction companies. However, in both cases, the primary R&D agents of nuclear power—the AEC in the United States and industry in West Germany—were to alter their roles as developers during this period and become the leading advocates.

Single Reactor Demonstration

On the surface, the demonstration strategies of the nations that had developed only a single reactor type was quite similar. Each had only one reactor to demonstrate, and the policy question was apparently one of priorities. The Canadian nuclear actors devoted the sum of their efforts to developing the heavy water cooled and moderated CANDU reactor; the French CEA concentrated its effort on the gas-graphite reactor while EdF lobbied for the more efficient LWR; and the British built and operated Magnox reactors while openly acknowledging that they were only interim power reactors.

Canada

The Canadian case is probably the most straightforward. In 1959, well before the completion of the Rolphton NPD reactor, the Canadian AECL announced plans to build a full-scale CANDU nuclear reactor at Douglas Point, Ontario.[66] Like the NPD facility, it would be built by AECL and turned over to Ontario Hydro as soon as it was declared operational.[67] Designed to generate 200 MW(e), it would be the first large-scale commercial-sized heavy water reactor.[68] Besides physical size, the biggest design difference between the NPD design and Douglas Point was that the latter's fuel bundles were to contain slightly enriched uranium to facilitate a faster start-up after a shutdown (see Table 5-5).

Even though the heavy water reactor seemed to be well on its way to general acceptance, the Douglas Point generating station raised two concerns. The first, which was relatively minor, was that fabrication problems in welding the joints and the seals in the heavy water transportation system delayed construction. The Douglas Point plant was due to open in 1964, but this as well

---

[66] Canadian nuclear technology at this time is summarized by W.B. Lewis, "Canada's Steps Toward Nuclear Power," in *Progress in Nuclear Energy-Economics* (London: Pergamon Press, 1959), vol. 2, pp. 227-288.

[67] See "Ontario Hydro's Nuclear Activities," p. 502, which states that the utility "agreed to purchase the $85 million plant from the AECL when it had been demonstrated a suitable addition to the power systems."

[68] A number of smaller heavy water reactors were in operation. *Nuclear Engineering* 11, 121 (June 1966), devotes the entire issue to heavy water reactors. Also see "Heavy-Water Reactors," in Robert L. Loftness, *Nuclear Power Plants* (New York: Van Nostrand, 1964), ch. 5, especially Table 5.1, "Chronology of Heavy-Water Reactors," p. 236, for descriptions of heavy water reactors in Norway, Germany, France, and the United States.

as other delays delayed the date until January 1967. The second problem, the cost of heavy water, was much more serious. A significant fraction of the cost of the CANDU reactor compared with the LWR is the price of the heavy water;[69] both Law and Hayward conservatively estimated that the D_2O cost may be as much as 15 to 20 percent of the total plant costs.[70] Canadian D_2O price projections in 1966 forecast that by the mid-1970s, the cost of producing heavy water would drop from $30 per pound to $14 per pound because of increased production.[71] This predicted decline was vitally important in the early 1960s because of two technical characteristics of the Canadian nuclear program. First, the CANDU reactor required large amounts of heavy water to operate—roughly one megagram (approximately a long ton) of D_2O per megawatt of new electricity capacity[72]—and second, Canada did not have sufficient domestic resources to produce the required amounts. Until 1967, Canada's heavy water was mostly purchased from the U.S. AEC's Savannah River Plant in South Carolina.[73] In other words, the CANDU reactor system was necessarily reliant on foreign suppliers. Therefore, AECL had to make certain that sufficient quantities of D_2O would be available before committing Canada to a large number of CANDU reactors at home as well as the export market they hoped to gain. The supply problem became so critical in the late 1960s that Douglas Point had to be shut down so that its heavy water could be transferred to the opening of the 500 MW(e) Pickering generating station.[74] Ontario Hydro was not able to reopen Douglas Point until a few years later when it was able to borrow D_2O from

---

69 McIntyre, "Natural-Uranium Heavy-Water Reactors," pp. 21-22, attempts to draw capital, fueling, and O&M cost comparisons between the CANDU and light water resources.

70 See C.A. Law, "Heavy Water Production," *Nuclear Engineering* 13, 145 (June 1968):510; and L.R. Haywood, "Heavy-Water Production," *Nuclear Engineering International* 19, 217 (June 1974):507.

71 See John O. Holt, "Canadian HWR: Economic Assessment," *Nuclear Engineering* 11, 121 (June 1966):461-463.

72 McIntyre, "National-Uranium Heavy-Water Reactors," p. 25.

73 Law, "Heavy Water Production," pp. 510-512.

74 This closure was particularly critical because part of the electricity generated by the Douglas Point facility was used to power a nearby D_20 plant. "3000 MW CANDU at Douglas Point," *Nuclear Engineering International* 14, 152 (January 1969):41.

Quebec's Gentilly station. To alleviate this problem, the Canadian AECL initiated a massive heavy water facility building program at Port Hawkesbury (Nova Scotia), Bruce (Ontario), Glace Bay (Nova Scotia), and La Prade (Quebec). The major contractors were two private firms, Deuterium of Canada and Canadian General Electric.

Thus, by 1964, Canada had made a major—although not irrevocable—commitment to the CANDU reactor system. Most of the difficult technological questions had been confronted by AECL technicians with NPD,[75] and with the exception of the heavy water supply problem, the future looked promising. However, the liberal subsidization policies of AECL suggested some hesitancy by Canadian industry to accept the CANDU system. The Canadian government had financed 75 percent of the Rolphton NPD facility, but private industry had financed only 5 percent. The dismal economics of heavy water production when the Port Hawkesbury plant went into operation in 1967 produced a steady financial drain on Canadian General Electric. CGE's condition was exacerbated by its AECL heavy water contract, which called for a declining price schedule that, as D_2O prices rose, ultimately forced CGE out of the market.[76] The brightest spot in the Canadian nuclear picture was the continued willingness of Ontario Hydro to participate in AECL's programs. Ontario Hydro provided 25 percent of the funding for the NPD, was the operating utility for the Douglas Point reactor, and in 1964, placed an order with AECL for two 500 MW(e) CANDU reactors to be built at Pickering. Ontario Hydro contributed "approximately 40 percent of the $272 million cost, the federal government 33 percent and the Ontario government 27 percent."[77] The utility was committed to repay both governments their contributions plus interest. This order was based upon the operational factors gleaned from the NPD reactor and

[75] For an account of some of the difficulties encountered with NPD reactor, see Eggleston, *Canada's Nuclear Story*, pp. 332-335.

[76] In 1975, CGE's contract with AECL required it to sell heavy water to the AECL for $17/lb while the open market rate was close to $50/lb. Rather than force CGE to absorb such losses, the AECL agreed to buy the Port Hawkesbury facility for $90 million and operate the plant itself.

[77] "Ontario Hydro's Nuclear Activities," p. 502.

an estimated electricity cost of slightly below 4 mills a kilowatt-hour, which compared favorably with a contemporary coal cost estimate of "about 4.0 mills."[78] Pickering was viewed as perhaps the critical facility in the Canadian nuclear RD&D and dissemination processes. As an Ontario Hydro executive noted:

> I do not believe I can overemphasize the importance to the Canadian nuclear industry of a favorable performance over the next few years at Pickering. The product of all our efforts is on trial. Other utilities at home and abroad are eagerly awaiting reports on the first unit of the first domestic commercial CANDU station.[79]

In summary, Canada's CANDU appeared to be a well-developed and established system by the end of the demonstration period. The major unanswered questions that would prove decisive to the Canadian nuclear reactor program were whether the other utilities would follow Ontario Hydro's leads and the involvement of private industry.

Great Britain

The United Kingdom found itself in a peculiar position in the mid-1960s. By most measures, the Magnox reactors had proven their worth and performed to expectations.[80] By 1971, England would have eleven prototype and full-size Magnox reactors with a capacity well over 5,000 MW(e)[81] (see Table 5-6). Substantial technical advances had been made during the design and construction of the Magnox series. For example, between Berkeley and Wylfa, the thermal efficiency of the stations increased from 24.6 to 31.4 percent, the capital costs decreased from $185/kw to $110, and the generating cost fell from 1.25 pence/KwHr to 0.7. Thus, by most accounts, it would appear that the British pro-

---

[78] The cost figures are cited by Eggleston, *Canada's Nuclear Story*, pp. 341-349.

[79] "Pickering—The First Year," *Nuclear Engineering International* 17, 193 (June 1972):475.

[80] See "Magnox Stations Lead in Electricity Production," *Nucleonics* 24, 3 (March 1966):48-50.

[81] See K.P. Gibbs and D.R.R. Fair, "The Magnox Stations: A Success Story," *Nucleonics* 24, 9 (September 1966):48-53. Both Gibbs and Fair were affiliated with the CEGB when this article was written.

totype phase had been neatly integrated with the demonstration phase and that the commercial acceptance of the Magnox reactor was imminent if not already a fact. However, a number of technical, economic, and political indicators strongly argued to the contrary.

First, it should be remembered that the U.K. AEA had never viewed the Magnox reactor as the British reactor that would be operated until the fast breeder was operational. British atomic scientists were especially conscious of the size and efficiency restraints imposed by using natural uranium fuel. With the availability of U-235 from the British separation plant at Capenhurst and the U.S. guarantee under the Atoms for Peace plan, enriched uranium was now an accessible commodity. Second, organized labor began to question if Britain should make a total commitment to nuclear power because of "the fear of unemployment in the coal mining industry. Where coal mining is the major industry, too rapid a drop in coal consumption . . . could create serious social problems."[82] Third, the spectre of the fuel shortages that motivated the earlier RD&D was less pressing. The British White Paper on the Nuclear Power Programme issued on June 20, 1960 reported that "Since 1957 coal has become plentiful and oil supply prospects have also improved. The need on fuel supply grounds for an immediate and sharp acceleration in the role of ordering nuclear capacity has therefore passed."[83] Fourth, the five industrial consortia were being hard pressed to survive economically because the U.K. nuclear market was too limited and restricted to justify the industrial capabilities that had originally been assembled in 1956; the tendering of one power station a year was inadequate to support all the consortia. To relieve the condition, the five consortia were reduced by 1962 to three. Nuclear Power Plant and AEI/JT were consolidated into the Nuclear Power Group (TNPG), each owning 50 percent; EE/BW/TW stayed intact, with each member owning one-third; GEC/SC and APC merged to organize the United Power Company (UPC), GEC being the principal

---

[82] Sir Rudolf Peierls, "Britain in the Atomic Age," *Bulletin of the Atomic Scientists* 26, 6 (June 1970):41.

[83] Quoted in "The Nuclear Power Programme," *Nuclear Engineering* 5, 50 (July 1960):316.

stockholder.[84] This situation worsened when the expected export market for the Magnox reactors did not materialize, especially in light of the predicted success of the American reactors in the foreign markets.[85] Fifth, CEGB and its chairman, Sir Christopher Hinton, were agitating for a more efficient, economic reactor. Sixth, there "was an economic crisis which forced a reduction in the rate of investment and brought an increase in the interest rate," thereby restricting the capital needed for a large atomic generating station.[86]

These pressures were addressed in another White Paper in April 1964, *The Second Nuclear Power Programme*, which found "that the economic advantages to be gained by the further development of the Magnox system were likely, in the circumstances now prevailing in the United Kingdom, to be smaller than those offered by other new systems."[87] This report suggested that the advanced gas reactor (AGR) would be the most likely candidate for the next generation of British reactors but called for an evaluation of "water-moderated reactor systems of proven design."[88] It defined the new English reactor goals as an additional 5,000 MW(e) during 1970-1975. The Windscale AGR, rated at 40 MW(e), had been operating at full capacity since February 1963, long enough to give AEA and CEGB confidence that the AGR was an acceptable reactor system and that Great Britain could safely proceed to a full-scale AGR.[89] As a contem-

---

84 See "Structure of the British Nuclear Industry," especially pp. 97-98. In 1963, perhaps out of sympathy, the two major British trade journals, *Nuclear Engineering* and *Nuclear Power*, merged "in line with the general consolidation of activities that has taken place throughout the nuclear industry over the last few years," *Nuclear Engineering* 8, 83 (April 1963):35.

85 Fear of the collapse of the export market was a persistent theme in the British reactor RD&D program, even given the early success. See two early *Nuclear Engineering* editorials, "Export Policy," 1, 1 (April 1956):2; and "Export Effort Inadequate," 3, 25 (April 1958):75-77. Doubts were again raised after the glow of the Japanese and Italian orders; see Editorial, "A Commercial Policy for the AEA," *Nuclear Engineering* 7, 77 (October 1962):381-382.

86 Perierls, "Britain in the Atomic Age," p. 41.

87 "The Second Nuclear Power Programme," *Nuclear Engineering* 9, 96 (May 1964):158.

88 Ibid., p. 158.

89 See "The Advanced Gas Cooled Reactor," *Nuclear Engineering* 6, 59 (April 1961):151-157.

porary British observer examined, "Our confidence springs both from the outstanding performance at Windscale and from our wide experience of the gas-cooled reactor—an experience that is not equalled by any other reactor system in the world. The A.G.R. is the most advanced system for nuclear power generation available now or in the immediately foreseeable future."[90]

To implement this new charter, CEGB requested bids for a 1,000 MW(e) advanced gas reactor to be built next to the Dungeness A power station. In February 1965, bids were received from APC (formerly the UPC until GEC/SC withdrew in early 1965) for a 1,200 MW(e) AGR but also from TNPG for a 1,050 MW(e) BWR and from EE/BW/TW for a 1,250 MW(e) PWR; the latter two would be built under American license obtained from GE and Westinghouse, respectively.

All three reactors were evaluated under U.K. operating conditions. This was an extremely important condition because in 1964 the New Jersey Central Power and Light Company had issued a report explaining its decision to purchase the 515 MW(e) Oyster Creek BWR (which could be expanded 25 percent to 640). The cost estimates, as low as $108/kw and below 4 mills/KwHr at full plant capacity, were reported as a major breakthrough in nuclear reactor costs. The Oyster Creek reactor design was thus viewed as a direct competitor to the AGR design supported by CEGB and AEA. AEA specifications were that a reactor have a plant life of twenty years and 75 percent load factor; the Oyster Creek analysis was based on thirty years and an 85 percent load factor. All tenders for the Dungeness B contract were adjusted to 1,050 MW(e) for comparison purposes. The Appraisal Board largely ignored the PWR tender, found the BWR design acceptable, but the AGR was the preferred choice. The report listed five reasons that underlay their selections: (1) higher thermal efficiency; (2) the AGR produced steam at temperatures and pressures similar to those from modern coal-fired generating stations, and thus the same turbines could be used, but the BWR produced saturated steam requiring specially fabricated turbine blades that could operate only at slower speeds; (3) the AGR offered

---

90 J.C.C. Stewart, "Frontiers of Technology: United Kingdom," *Nuclear Engineering* 9, 100 (September 1964):311.

higher reactor availability because of on-line refueling; (4) greater safety; (5) and finally, the AGR offered a greater development potential.[91] Table 6-1 reproduces the Appraisal Board's financial comparison of the two plants.

Although the AGR was the clear economic choice, critics of the Appraisal Board pointed out that the BWR was scheduled to cost "80 percent higher than the cost of the Oyster Creek BWR plant if it reached the full capacity of its turbines, and nearly 50 percent higher if the Oyster Creek reactor did not exceed its initial capacity, which no one regards as a probability."[92] In addition to the five stated reasons, the possibility that the AGR would revitalize the dormant Capenhurst separation plant and be a valuable export commodity were thought to have influenced the decision.[93] In any case, the decision was debatable. An editorial in *Nuclear Engineering* referred to it as a "specious selection," largely because of the greater growth potential of the LWR and its more immediate availability.[94] On balance, the choice prob-

Table 6-1. CEGB Estimate of Generating Costs for AGR and BWR (£ per KW)
Adjusted for Operations at 1200 MW(e)

	AGR	*BWR*
Total capital cost	78.40	70.86
Initial fuel charge	15.44	17.26
Final fuel credit	(1.58)	(1.68)
Replacement fuel	25.92	30.46
Availability adjustment	–	8.97
Other work costs	6.09	7.21
Insurance	3.41	3.41
Total	127.68	136.49
p/KwHr	0.457	0.489

Source: CEGB, "An Appraisal of . . . Dungness 'B', p. 145.

---

91 Central Electricity Generating Board, *An Appraisal of the Technical and Economic Aspects of Dungeness "B"*; reprinted in *Nuclear Engineering* 10, 112 (September 1965):144-146.

92 Burn, *The Political Economy of Nuclear Energy*, p. 24.

93 Victor K. McElheny, "Decisions on Nuclear Power," *Science* 149, 3682 (July 23, 1965):408-409.

94 Editorial, "Specious Selection," *Nuclear Engineering* 10, 109 (June 1965):108-109. The journal's position was forecast by their earlier evaluation of

ably owed a great deal to the fact that the AGR was British designed and manufactured. In 1965, a 1,200 MW(e) AGR for Dungeness B was ordered, and four additional large AGR plants were ordered by 1971. The English AEA and nationalized utility had clearly decided to exercise the AGR option as the second stage of its nuclear reactor development and commercialization program.

Almost immediately after the decision, there were numerous indications that the second stage was in serious trouble. In 1967, less than two years after the AGR-BWR appraisal, revised AGR "estimates indicated a probable [KwHr] cost of 0.52p, an increase of over 10 percent."[95] Delays in the AGR construction schedule began to mount, mostly because of inadequate initial designs necessitating major changes as the construction progressed. New technical information developed during the construction period identified corrosion problems that would either limit the life of the boilers or require them to be operated at reduced temperatures—that is, reduced plant efficiency. For example, the 9 percent chrome steel used in the boiler tubes was found to have corrosion tendencies after an initial period of low corrosion.

At the same time, the British nuclear industry was experiencing yet another painful restructuring. By 1962, the original five consortia had been reduced to three because "of the fluctuating fortunes of the nascent nuclear industry and more particularly the dearth of orders and the rapid growth in the size and capacity of nuclear power stations (thus causing a reduction in the number of individual power stations required to arrive at a given nuclear capacity)."[96] In 1965, the GEC/SC components of UPC had withdrawn, leaving only APC, which had been awarded the Dungeness B AGR contract. Even after this reduction, the Select Committee of Science and Technology recommended in October 1967 that there were not enough prospective orders to support three firms, that the government should become involved with these firms in integrating their nuclear construction capabilities,

---

the government's second nuclear program, which the editors characterized as "The Anti-Programme," *Nuclear Engineering* 9, 96 (May 1964):151-152.

[95] Jensen, *Nuclear Power*, p. 109.

[96] Ibid., p. 112.

and that the projected nuclear construction industry would be most efficiently operated by a single large firm.[97] Although the recommendation was not accepted, the Industrial Reorganization Corporation (IRC), owned by the government, was formed to help consolidate the industry. To this end, in September 1968, it was announced that there would only be two large British nuclear construction consortia, composed of a partnership between private industry and government. British Nuclear Design and Construction (BNDC)[98] was formed in October 1968 and the Nuclear Power Group (TNPG) the following January. The government, through the AEA and the IRC, owned 46 percent of BNDC and almost a third of TNPG. BNDC was given the defaulted Dungeness B contract.

The British nuclear situation continued to deteriorate to such a point that the CEGB ordered an oil-fired plant, Sizewell B, in place of the AGR that had been originally planned for this location. Thus, by 1972, both the rising costs of the AGR and the declining fortunes of the British nuclear industry implied that the AGR demonstration phase had been a great deal less successful than originally predicted. In a sense, the demonstration period had been successful in that it revealed the weaknesses of the AGR reactor. However, this was a Pyrrhic victory that might have been avoided had more discriminating analysis been performed prior to the AGR selection or had the technological and economic uncertainties of reactor development not been obscured by nuclear patriotism.

Burn argues that the British choice was dictated by two features. First, the "narrow development front" effectively presented the British decisionmakers with a Hobson's choice. Once the decision was made in favor of nuclear power, the lack of enriched uranium and alternative reactor types all but demanded the Magnox, and then the follow-on AGR types. The reason behind this single choice, Burn asserts, was that AEA monopolized the R&D and nuclear decisionmaking process, excluding manufacturers and even occasionally CEGB from its delibera-

[97] The report of the Select Committee is excerpted in *Nuclear Engineering* 18, 140 (January 1968):30.

[98] Initially called Babcock (and Wilcox) English Electric Nuclear (BEEN).

tions. Burn writes that "unlike the AEC, [the AEA] was not prepared to let other manufacturers 'come in and pick out' reactors to develop 'as they see fit'."[99] Only the forceful intervention of the exchequer persuaded AEA that CEGB should be included in, or at least consulted on, the early Magnox decisions. The second reason lies in the British nationalized electricity distribution system. CEGB was the sole purchaser for British atomic reactors except for Scotland. One or maybe two clients were not able to generate a sufficiently varied set of demands that would have resulted in a number of different reactors systems being developed. A third possible reason was that AEA and CEGB guaranteed each consortium member that it would receive contracts for nuclear generating stations. Therefore, there was a greatly reduced level of industrial competition among the members.[100] A strong Tory, Burn summarized his argument:

> In contrast to the American organization, the British is highly personal. The AEA has the monopoly of choosing and developing new reactor systems: the CEGB is sole buyer of nuclear plants in England and Wales. The members of both are appointed by ministers, who are nominally responsible for the efficient operation of two monopolies. The nuclear plants bought are not chosen solely on commercial criteria . . . Private firms were given no creative part in the early development of the nuclear industry . . . The advantages the Americans have derived from relying increasingly in all stages on private industry . . . have all been sacrificed in the development leading up to the completion of the prototype.[101]

The details of Burn's attack are subject to debate.[102] CEGB is hardly the handmaiden of AEA; CEGB lobbying urged AEA to abandon future Magnox reactors and, furthermore, demanded that the LWR be included in the list of alternatives. Although the involvement of British industry in the nuclear reactor RD&D processes was certainly less than in the United States or West Germany, it was not as isolated as Burn contends. Schattmann

---

[99] Burn, *The Political Economy of Nuclear Energy*, p. 87.

[100] For an early indication of this problem, see Editorial, "Commercial Power Stations," *Nuclear Engineering* 2, 10 (January 1957):39.

[101] Burn, *The Political Economy of Nuclear Energy*, p. 113.

[102] An assessment of the Burn monograph with qualifying comments by AEA personnel is provided by John Walsh, "Nuclear Energy: New Study Assails British Program," *Science* 156, 3780 (June 9, 1967):1350-1352.

wrote after the opening of Calder Hall, "Perhaps the most striking feature of the rapid rate of progress in using atomic energy for peaceful purposes on a large scale in Britain is the fact that industry came in at an early stage to undertake commercial development of reactor systems designed by the U.K. Atomic Energy Authority."[103] Still, even if his particular agreements are suspect, Burn's analysis correctly focuses on the root causes of England's nuclear difficulties, the single reactor research strategy adopted immediately after World War II and its refusal to accept foreign reactor technology. Once channeled into that strategy, AEA chose to become the sole developer in a constrained market, thus inhibiting the influx of new technologies and designs.[104] The closed alternative set was even more restricted by the decision to support a continuing series of British nuclear reactors rather than adopt the German (and later French) strategy of abandoning the current generation of reactors to American reactor designs and concentrating on the future reactor generations, such as the HTGR.

Therefore, at the same period in the developmental cycle as the Germans, Americans, Soviets, and the Canadians were weighing the possibilities of widespread dissemination of their reactor designs, the British nuclear RD&D actors found themselves burdened with rapidly escalating cost predictions, stubborn and difficult technical problems, and a decaying nuclear industry. The English nuclear policy question in 1972 was not one of reactor dissemination, but the much earlier question of which reactor should be developed. In effect, the British reactor demonstration phase had retrogressed to the prototype development phase.

France

The demonstration period in France was technologically characterized by the cautious French demonstration of the gas-

---

103 Schattmann, "Britain's Nuclear Power Program," p. 1489.

104 This is documented at length by Edward F. Wonder, "Decisionmaking and Reorganization of the British Nuclear Power Industry," *Research Policy* 5, 3 (July 1976):240-268.

graphite reactor. A few new CEA reactor innovations were developed during this period, such as a new annular high performance fuel element (INCA) that was introduced at Bugey. Chinon-3, a Magnox fueled reactor rated at 500 MW(e), became operational in August 1967; St. Laurent-1 entered operations two years later (see Table 5-7). Both were EdF facilities. The latter plant design was reputed to have "attained technical and economic levels at which competitiveness with fossil fueled systems can emerge."[105] Technically, therefore, the French demonstration period was marked by a steady refinement and expansion of the earlier French gas-graphite reactor designs.

Institutionally, the trends that surfaced during the previous period began to accelerate. Specifically, EdF gained additional leverage as a leading nuclear actor in the French context, and French private industry's responsibilities and participation became more pronounced. The responsibility of the St. Laurent series of reactors and Bugey-1 belonged entirely to EdF rather than being joint CEA/EdF projects. At the same time, French industry became more actively involved in the national nuclear program as opposed to its earlier supportive (or supplier) role.[106] In 1966, corporate mergers reduced the number of nuclear consortia from five to three, GAAA, Alsthom, and SOCIA. For the first time, CEA surrendered its monopoly in the design and fabrication of fuel elements to private industry.[107] Similarly, EdF, which had been even more relucant to utilize private industry than CEA, permitted a consortium of French industrial firms to serve as the industrial architect for the Franco-Spanish gas-graphite reactor built at Vandellos, a role it had previously reserved for itself. Because of its increased reponsibilities, French industry began to have an effect on the French reactor design. For example, the two St. Laurent reactors utilized steel rather than prestressed concrete pressure vessels, a change motivated in part

---

105 Robert Hirsch, "French Nuclear Policy," *Nuclear Engineering* 11, 127 (December 1966):938.

106 See Roger Julia, "French Industry Faces the Challenge of Nuclear Engineering," *Nuclear Engineering* 11, 127 (December 1966):943-944.

107 For a retrospective view, see G.Y. Kervern, "Industrial Production of Nuclear Grade Materials," *Nuclear Engineering International* 17, 196 (September 1972):707-709.

by the industrial observation that the size and weight of the concrete vessel vitiated whatever export opportunities existed for the gas-graphite reactors.

As Nau notes, during the demonstration period in France, "the first commercial sale of a light water reactor in the United States in 1964 generated initial pressures for a reorientation of French reactor and industrial policies."[108] This refers to the same Oyster Creek announcement that had caused the nuclear brouhaha in Britain and resulted in the comparative evaluation between the AGR and the BWR before AEA chose the British reactor for Dungeness B. The CEA reaction was much more xenophobic. Basically, CEA acted as if Oyster Creek were completely irrelevant. CEA maintained its active research program on the gas-graphite design (with long-term emphasis on the breeder), and in late 1964, the PEON group announced the continued French reliance on the gas-graphite reactor, although suggesting other reactor types be investigated.[109]

In contrast, EdF and parts of the French nuclear industry started to lobby intensely for the French adoption of the American LWR technology. The loss of the Fessenheim 1 Franco-German reactor, which occurred when Siemens refused to accept the French reactor, alerted French industry that the export value of their reactor was becoming limited. For the potential reactor export market, it was increasingly clear that the "French reactors . . . had capital costs so high that they were noncompetitive as export reactors . . . The French saw themselves becoming technologically isolated and foreclosed from competing successfully for sales abroad."[110] CEA authorities commented at the time that the competition was not entirely based on reactor technology and that the "United States and Canada have . . . practiced a policy of credit which involves commitments that it would be difficult for France to envisage."[111] Still, it was recognized that success in foreign sales was more than a matter of credit terms and that, in matters of nuclear technology, the

---

108 Nau, *National Politics and International Technology*, p. 83.

109 *PEON*, Reports, Tome I, p. 14.

110 John Walsh, "Nuclear Power: France Forges Ahead on Ambitious Plan Despite Critics," *Science* 193, 4250 (July 23, 1976):340.

111 Hirsch, "French Nuclear Policy," p. 939.

French gas-graphite reactor did not appear viable in the export market.

Within the domestic market, there was increasing evidence of the relative economics of the LWR over the gas-graphite reactors. This was reinforced by technical problems experienced in the French reactor system, especially in 1966 during the construction of the heat exchange system in the Marcoule-3 facility. The repair costs for this plant were close to 10 percent of the total plant cost.

By the middle of 1967, the French nuclear community was sorely at odds over the French government's choice of reactor. An article in *Le Monde* characterized CEA and EdF as "deeply divided" and that relations between the two "never ceased to be tense" owing to each one's desire to affirm its supremacy in nuclear reactors.[112] EdF cited economics of operation as its principal reason for supporting an LWR. French industry held that the gas-graphite reactor was particularly difficult to export because of its size. Its members argued that its freedom from the necessity of enriched uranium was its strongest selling point, but given that the St. Laurent-1 reactor was as efficient a gas-graphite reactor as French technicians could produce, the export potential was apparently very restricted.

Subscribing to the tradition of French nuclear independence—or more precisely the refusal to become dependent on American technology—CEA deflected the pressures being brought to bear in favor of the LWR. Rather, it chose to investigate natural uranium heavy water cooled reactors (EL4) and the British AGR developments. CEA's logic was clear and self-serving for both itself and France. If CEA could resist the American nuclear technology by either improving the gas-graphite reactor or developing a new reactor type, it would preserve French independence from American industry as well as continue CEA's hegemony in the French nuclear RD&D programs. Any LWR inroads into the French nuclear community would mean a further diminution in CEA's influence. In short, the French situation could be viewed as an institutional struggle

---

[112] The undated article by Nicolas Vichney was reported in the news column of *Nuclear Engineering* 12, 130 (March 1967):150.

between EdF and CEA, "with the EdF tending to assert its duty to cut generating costs even at the price of using American techniques and fuel, while the CEA, directly under the Premier, was advocating a different energy policy."[113]

For these reasons, in late 1967 French nuclear planners began to cast around for alternative technologies. In the spirit of the Concorde aircraft development, French Science Minister Maurice Schumann discussed "much closer links with Britain, with a view to increasing to a maximum the co-operation between the CEA, EdF and their equivalents across the Channel on gas cooled graphite reactors." This joint research could have focused on either the Magnox or the AGR types, but there was little mistaking Schumann's political intention when he said that these designs were

> the only big nuclear technique competitive with the transatlantic types. Franco-British cooperation could have a decisive influence on the future of European projects like the OECD Dragon and the Euratom high temperature THTR type developed at Jülich. This shows that the course selected, far from being that of the past and a closing in on oneself, is in reality the only one enabling Europe to meet U.S. competition.[114]

He closed by stating that the French could not build American reactors in France that would render her dependent upon U.S. enriched U-235 as a source of fuel.

Besides the Magnox and AGR options, which represented a continuation of their existing technologies, French nuclear planners, particularly in EdF, considered the BWR and the PWR. The BWR European license was held by AEG; furthermore, GE did not seem interested in granting a license to a French concern.[115] The PWR situation was more favorable. CEA had enjoyed some success since 1964 with a land prototype PWR designed for the atomic French submarine. Therefore, adoption of the PWR would permit CEA to use some domestic technology. Finally, CEA in-

---

113 "Changing Trends in France," *Nuclear Engineering* 12, 139 (December 1967):882.

114 His speech was reported in "GCR Collaboration," *Nuclear Engineering* 13, 131 (February 1968):81.

115 See "Changing Trends in France," p. 882.

itiated a series of technical exchanges with the Canadian AECL for heavy water power reactors.

Following a cabinet-level review of French nucler reactor policy, a decision was made in December 1967 by the deGaulle government that France would continue its reliance on the gas-graphite reactor. The argument was made that the acceptance of the LWR type "would have involved either an unacceptable degree of reliance on American supplies or the construction of a European diffusion plant—such a plant would not only be outside strictly French control but it would require detailed, protracted and probably unwelcome negotiations with a number of other European countries."[116] However, the force of the industrial and EdF arguments was not completely ignored. CEA agreed to build a second PWR plant in conjunction with Belgium, the 725 MW(e) SEMO facility in Tihange, Belgium, which would be built by Framatome. The French government appeared to place significant emphasis on the SEMO reactor. Even after the French riots of 1968 and the subsequent reduction of the national budget, which forced cancellation of plans for gas-graphite reactors scheduled under the fifth Nuclear Program (1966-1970) and the reduction of CEA staff and scientists, the Tihange plant proceeded on schedule.

French industries, perhaps interpreting SEMO as an early indication of a possible government switch to the LWR, began to maneuver to obtain licenses with the U.S. nuclear companies. New consortia were formed; Alsthom obtained a GE license and merged into SOGERCA while Framatome added the Westinghouse PWR licenses and its Chooz and Tihange experience to SOCIA.

Thus, by the middle of 1968, the development of the nuclear reactor in France was more affected by bureaucratic conflict than by technical difficulties over the chosen reactor. CEA insisted on the continued development and demonstration of the French gas-graphite reactor. The nationalized French electric utility created an important counterweight. EdF largely based its reactor preference on a set of economic objectives or criteria that differed from the protectionist tone of the CEA positions. Once it decided

---

116 Jensen, *Nuclear Power*, p. 128.

to become involved in the generation of electricity by nuclear reactors, EdF had the political influence to affect the French reactor program.[117] An economic objective function was also adopted by the emerging French nuclear industry, especially as it saw the West German market being lost to the United States.

The demonstration of the French gas-graphite reactor technology continued during this period, but the main decision about the widespread acceptance and dissemination of the reactor was still in abeyance. French policymakers were being forced into a decision that CEA wished to avoid, a choice between the economically problematic French reactor and an increasingly cost-effective U.S. reactor that had been developed entirely outside of France.

French and British Comparison

The French and British were both forced to make a fundamental nuclear reactor decision on the viability of their indigenous designs. The choice was more than a technical assessment of the competing nuclear technologies and economies of construction and operation. The debate was highly biased by partisan sentiments and national prestige, many of the same factors that originally induced France and Britain to build nuclear power reactors.

The existence of nationalized utilities in each nation confounded the issue. The nationalization of the electricity industry both gave the governments a valuable policy lever and protected the industry from external economic penetration, which generally meant the United States. Scheinman comments:

> In countries in which electricity production, the principal consumer of peaceful nuclear technology, is nationalized, government may play a role not only in the selection of power sources, reactor types, and production strategy but also in the structure of the nuclear industry itself. Thus, the governments of a number of West European countries have intervened in the electrical equipment industry to help create powerful national enterprises capable of effectively competing with external suppliers. It can be hypothesized that these efforts are a

117 See Bupp and Derian, *Light Water*, pp. 60-69.

> defensive response to the perceived risk of extensive penetration of national energy markets by external suppliers.[118]

This governmental policy of protecting its industries was highly successful in France and England, but an ironic twist in the execution of this policy undermined both national nuclear programs. The nationalized electric utilities and the structured nuclear industries were able to resist foreign (especially American) business and technology encroachments but they also grew to become major participants—rather than recipients—in their country's nuclear decisionmaking processes. Given their differing objective functions, it was not surprising that they became vocal and later powerful opponents of national nuclear priorities as expressed by the nuclear research agencies. In effect, the nationalized electric utilities and the nuclear industries, which had been created and fostered partially to resist U.S. economic or technological penetration, became the proponents and vehicles of precisely the forces they were created to resist.[119] Their arguments were more economic than nationalistic. They were therefore particularly difficult to ignore as the French and British entered periods of economic distress in the late 1960s and early 1970s.

Summary

Each nation used the demonstration period to determine if large-scale nuclear reactors were operationally and economically feasible methods of producing electricity. Each nation conducted and operated the power reactor(s) it had developed in the previous prototype development phase but on a larger scale. As Scheinman notes, nuclear research cannot be clearly separated from development, "In order to develop a competitive reactor it is necessary to build a series of prototype or demonstration reac-

---

118 Scheinman, "Security and a Transnational System," p. 282.

119 The manner in which components or new members of a coalition can alter the institution's objectives is a concern of Albert O. Hirschman, *Exit, Voice, and Loyalty* (Cambridge, Mass.: Harvard University Press, 1970).

tors."[120] By the end of the demonstration period, each nation had sufficient information to decide whether it should progress to the widespread dissemination of its chosen nuclear reactor. However, no case was cut and dried, no decision appeared to be easy. France and Great Britain were confronted with increasingly uneconomic reactors or accepting foreign technology with onerous costs. The Soviet Union had demonstrated the operational capabilities of two reactor types and was faced with the decision as to which one (or both) should be disseminated, if any. The choice would be made by the Council of Ministers. Canada and the United States faced much the same choice. Each had demonstrated the operational capabilities of a reactor (or reactors in the case of the United States), and the remaining question was whether the utility companies and nuclear industries could agree on the commercialization of nuclear reactors, more specifically, what conditions would be necessary to deliver a demonstrated technology to the consumer. A similar dilemma faced the nuclear industry in West Germany with the interesting amendment that German industry and utilities had to choose among foreign reactor technologies. The main difference between the Soviets and these last three nations is that the Soviet dissemination decision would be a political decision made by the government, while the American, German, and Canadian decisions would be shared among a wider range of participants. Although the Soviets would base their nuclear decision partly on economic factors, the Western nations' choices would be primarly determined by the forces in the economic marketplace.

The demonstration strategies employed by the different nations were not markedly different in intent. However, the end results were as varied as the strategies that brought these nations to this stage. Decisions that were made during the earliest periods of power reactor R&D were having a major effect more than twenty years later and would be further manifested in the wide dissemination or commercialization of full-sized power reactors.

---

[120] Scheinman, "Security and a Transnational System," p. 282.

7

The Dissemination of Nuclear Power Reactors

Definitions

The final stage of the RD&D process is the dissemination or commercialization of the product, in this case, nuclear power reactors. For now, the dissemination or commercialization stage may be defined as beginning with the national decision made either by the government, the electricity producers, reactor construction companies, or some combination of these, to build and operate (or not) a significant number of large-scale nuclear power reactors. The number of reactors that could be considered significant is related to the national context; the "significant" number of reactors in Canada would be insignificant in the American context. "Large-scale" refers to reactors that are rated at a minimum of 400 MW(e)—large enough that economies of operation characteristic of nuclear power stations would be realized.

The commercialization phase of nuclear power reactors may be viewed as open-ended. As long as slow fission reactors continue to be built and operated for the generation of electricity, the dissemination progress is continuing. By this definition, the termination of the dissemination stage would occur only when an alternative mode of electricity production becomes the dominant power source—in other words, when slow fission nuclear reactors are no longer built. Present purposes are better served by arbitrarily terminating this view of the dissemination stage when the nuclear industry either firmly accepts or rejects the nuclear reactor or reactors that have been developed and demonstrated to that point.

An example would be useful to illustrate this distinction. After 1967, it was obvious that American industry and utilities had accepted the LWR; the decision to commercialize the BWR and the PWR had been taken and the reactors were performing approximately as expected. Commercialization had occurred. But a longer term view or projection of the U.S. nuclear energy market might reveal that the dissemination of the LWR was not as "successful" as originally thought and that the safety and disposal questions, perhaps glossed over at first, were to present conclusive evidence that the U.S. electricity industry should have continued its reliance on fossil-fuel power plants until alternative energy resources were available.

The phrase "dissemination" of nuclear power reactors has been selected in preference to "commercialization" because of the difference in the market structure among the surveyed nations. The lack of a traditional economic market structure in the Soviet Union renders the accepted concept of "commercialization" worthless in the Soviet context. To a lesser degree, the nationalized electricity utilities and the carefully structured nuclear industries of France and England are resistant to the "commercialization" process as it operates in the United States and West Germany.

The dissemination or commercialization of nuclear power reactors follows the reactor research, development, and demonstration (RD&D) stages. It can be conceived to be either the most important or the least important in that it represents the final acceptance or rejection of a research process, the ultimate evaluation of the RD&D process. Research and development can carry a system only to a certain point, after which it must be accepted or rejected on its demonstrated merits; potential is no longer adequate. In this sense, the dissemination period is the most important of the three stages. The fundamental RD&D was largely completed, the prototype model performed adequately, a demonstration reactor was evaluated, and a final question remained. Did the end result warrant widespread dissemination; and, if so, could it be reasonably expected to compete and succeed on its own intrinsic merits or could it exist only with large government subsidies? If the last requirement were applicable, probably additional R&D would be necessary. The general reasons why the reactors were disseminated were political (e.g., a matter of national prestige), economic (e.g., they performed a needed service at a lower cost or more efficiently than contending alternatives), institutional (e.g., the R&D acquired a momentum that carried the reactor into a general acceptance), or a combination of these. For the general analytic process, however, the specific reasons underpinning the choice are unimportant when compared to the fact that the decision was made to accept or reject the product.

This stage could also be the least important because it merely represents the culmination of the entire research processes. That is, by the time a technical system reaches the final stage, it is so much a product of all that has gone before this time or development that the critical decision is a prisoner of prior events. York explains the conditions that led to the dissemination of the MIRV missile warheads: "The decision to deploy MIRVs was made all but inevitable by the decision to develop them."[1] In this perspective, the dissemination stage is just a formality, an acknowledgment and reification of all that has gone before.

Neither of the perspectives is entirely correct or completely wrong. They serve to bound the place and importance of the dissemination period. As might be expected, the six national experiences covered may be placed on various points of this line. For the United States, the RD&D on nuclear reactors was not conclusive; decisions made during the dissemination stage crucially affected the commercialization of the LWR. For different reasons, the dissemination period in France and England

---

[1] Herbert York, "Multiple Warhead Missiles," *Scientific American* 229, 5 (November 1973):26.

could hardly be described as mere extensions of the earlier periods. To various degrees, the dissemination stages in the Soviet Union, Canada, and West Germany can be treated as the linear culmination of the technology and decisions of the earlier RD&D periods.

This chapter, as were Chapters 5 and 6, is organized in terms of a rough dichotomy, the decision(s) within each nation to accept or reject the developed nuclear reactor. The nations that chose to disseminate their preferred indigeous reactor designs were the United States, the Soviet Union, and Canada. West Germany's situation is slightly different because their LWR was not a domestic design. The nations that chose not to disseminate their existing reactor designs were Great Britain and France. The implications of their rejections and how each decided to amend its nuclear programs were marked.

The general reasons for the acceptance of a reactor include economic incentives (especially in light of alternative power resources), the proven operational capabilities of power reactors, and some measure of national prestige. As will be seen, the first two are the most important, but the third cannot be dismissed. Jensen suggests, "But for the fear of being outdistanced in the race for the development of civilian nuclear technology, it is probable that the initial Soviet atomic power station construction programme would have been even smaller and less ambitious than was actually the case."[2] Similar cases can be made for the other national programs. The reasons underlying the rejection of reactor types at this stage are much the same. A reactor system might finally be recognized as uneconomical and presenting too many technical difficulties, or even as being detrimental to the national image. An example of the last point is France, whose government considered the penetration of the U.S. LWR into Europe as a national affront to French science, the French gas-graphite design, and European unity. But rather than continue to suffer in technical comparison and abandon the valuable reactor export market, France chose to minimize the losses to its prestige by rejecting the gas-graphite reactor, adopting the U.S. LWR, and placing the full weight of its nuclear scientific

---

[2] Jensen, *Nuclear Power*, p. 8.2.

establishment behind the development of the fast breeder. The reasons behind a nation's decision are closely analogous to those that were marshalled to initiate these nations' nuclear program.

Reactor Acceptance

United States

In 1968, John F. Hogerton, a noted commentator on American atomic energy, commented on the arrival of nuclear power:

> Nuclear power, like the boy next door, seems to have grown up overnight. That it has indeed come of age is incontrovertible. For two years running it has accounted for nearly half of all the new power-generating capacity ordered by U.S. utilities. More than 50 large nuclear generating units, representing a financial commitment of $7 billion, are now under construction or awaiting construction permits. Their combined capacity exceeds 40 million kilowatts, which is as much electrical plant as the U.S. had in service going into World War II.
>
> That maturity came quickly is also incontrovertible. . . . It is safe to say that no one, not even the most optimistic reactor manufacturer, expected so rapid or decisive a market breakthrough.[3]

Hogerton's last sentence was an accurate assessment of nuclear reactors and the American market in the early 1960s. Before mid-1963, the U.S. AEC and nuclear industry had expended considerable resources developing and demonstrating a number of nuclear reactor types but only three LWR reactors rated over 140 MW(e) had been built—Dresden 1, Yankee, and Indian Point 1.[4] Only the Haddam PWR, rated at 575 MW(e), had been ordered on a commercial scale. The American nuclear industry and public utilities appeared to be on the verge of accepting nuclear power but unwilling to take the final step. That step was presciently taken in late 1963 when three large power reactor orders were announced: New Jersey Central Power and Light ordered the Oyster Creek reactor, a 515 MW(e) BWR, from

---

[3] Hogerton, "The Arrival of Nuclear Power," p. 21.

[4] See Allen, *Nuclear Reactors for Generating Electricity*, for details on the different reactor developments and AEC programs.

General Electric, with the capacity to be expanded to 640 MW(e); the Niagara Mohawk Power ordered a BWR plant for Nine Mile Point, also from GE, that was rated at 500 MW(e) (later expanded to 620 MW[e]); and Southern California Edison ordered its San Onofre 1 reactor, a 430 MW(e) PWR, from Westinghouse. Not only was the size of these orders remarkable, but also the low costs of construction and electricity that was promised. President Lyndon Johnson referred to these reductions as an "economic breakthrough in nuclear power."

To appreciate the importance of these orders, one should understand that at the time the American coal industry was mounting a concerted strategy to reduce its prices and the cost of transporting coal to the generating station as a means of heading off the potential nuclear threat. For example, before the Oyster Creek announcement, Jersey Central had been purchasing coal at a delivered price of 30-31 cents per million BTUs of energy content. To compete for the Oyster Creek facility, the offered price of coal was lowered to 26 cents. Nuclear energy was hardly the sole bidder in the energy market.

In February 1964, Jersey Central released their comparative economic analysis of the Oyster Creek facility. The assessment evaluated three alternatives: a nuclear reactor power plant, a coal-fired plant of comparable size at Oyster Creek, and long distance transmission of electricity generated from a power plant located in a coal mining area. The economics of the comparison argued that the nuclear facility would be the least expensive. If the Oyster Creek BWR were operated at its nominal 515 MW(e) capacity, the price of coal would have had to have been slightly below 26 cents to be competitive; if the BWR were operated at its full 640 MW(e) capacity, the coal price would have had to have been below 20 cents, a price available only to plants in coal mining areas. The costs of the plant per kilowatt were far below what the AEC had predicted in its 1962 report to President Kennedy.

The economy of the plants would depend much on how soon the higher capacity was reached. But even at the lower guaranteed capacity (515 MW) the projected cost of the Oyster Creek plant, $134 per KW, was nearly 25 percent less than the cost forecast by the AEC. At 640 MW

> it was slightly below $108 per KW . . . a drop of almost 40 percent below the 1962 forecast.[5]

The operating cost of the plant at full capacity was 3.5 mills per KwHr.

The GE contract with Jersey Central was a critical component of the economic evaluation and one of the main reasons for the order. GE agreed to build the plant on a "turnkey" basis (i.e., the plant would be turned over to Jersey Central in operating condition) on a fixed price contract for $68.9 million. "If the Oyster Creek findings were at all indicative, nuclear power had arrived. The question was, how indicative was Oyster Creek?"[6] The question provoked a major debate, especially in light of Niagara Mohawk's statement that the Nine Mile Point reactor, with approximately the same capacity as Oyster Creek, was—at $90.2 million, $145/Kw, and 6.67 mills/KwHr—considerably more expensive.[7] GE's explanation of their pricing strategy was cogent and persuasive.

> They had reached the stage when they thought they could begin to make their enormous investment in nuclear power—possibly over $150 million—remunerative. Their BWRs were now proved and reliable, and several utilities had experience in using them. If their price could be brought near to that of a conventional coal-fired power station, many utilities would be ready to buy one. If a sufficient number of BWRs was sold then the price could be brought close to that of a coal-based plant. The price for Oyster Creek was thus set at a level which would at the best give a small profit if this was an isolated sale, but a satisfactory profit if, say, six were sold. The sale of several sets which in many respects were identical would reduce appreciably the cost of making the plant and of fabricating fuel elements.[8]

---

[5] Burn, *The Political Economy of Nuclear Energy*, p. 24.

[6] Hogerton, "The Arrival of Nuclear Power," p. 27.

[7] The price discrepancy between the Oyster Creek and Nine Mile Point GE BWRs is discussed and largely reconciled by Burn, *The Political Economy of Nuclear Energy*, App. II, pp. 140-141. Quoting a Niagara Mohawk vice president, Burn states that the utility deliberately used a more conservative accounting approach, which "probably put the cost of Nine Mile Point 10 percent higher than it would ultimately be." Land, land preparation, and water supply were close to $6 million greater for the Nine Mile Point BWR than for Oyster Creek.

[8] Paraphrased from a letter from Y.F. Young of GE to the JCAE, ibid., p. 25.

There was little consensus on the validity of GE's corporate decision, but four fairly intangible factors appeared to operate in favor of GE and the nuclear construction industry as a whole. First, the coal industry was already filling a record number of orders and its leaders might have calculated that they could safely afford to lose some contracts to nuclear power reactors. In addition, coal transportation costs—which accounted for roughly a fifth of the total cost of power generation in a coal-fired station—had reached a point beyond which further reductions were infeasible.

Second, there seemed to be "a growing conviction among utilities that nuclear power would be the way of the future."[9] Utility presidents might opt for a nuclear facility to provide their personnel with training in the new technology as well as appearing to be in the technological vanguard. This was part of the stated rationale that led Duquesne to sponsor the construction and later operation of the Shippingport PWR.[10]

Third, the electric utilities were initiating a series of interconnecting power grids between their electric networks, which encouraged them to build larger, more cost-effective power plants than they might normally do under a single-system basis. Nuclear power was considered to be characteristically more economic the larger the plant. In addition, the utilities were experiencing some downtime difficulties with their newest and largest conventional power plants. Both of these factors pointed advantageously to nuclear power reactors. Finally, the more cognizant utility executives were beginning to recognize the environment and air pollution movements and looked toward nuclear energy as a possible acceptable alternative.

Nevertheless, the general acceptance of power reactors as demonstrated by a widespread decision to purchase reactors was in suspension during the next eighteen months, with the notable exception of Commonwealth Edison's purchase of Dresden 2, a 715 to 793 MW(e) BWR from GE. "The turnkey contract price was 'not to exceed' $76.8 million, but there were arrangements for it to be reduced if the ultimate ('stretched') capacity was not

---

[9] Hogerton, "The Arrival of Nuclear Power," p. 29.
[10] Hewlett and Duncan, *Nuclear Navy*, p. 238.

reached in a specific time."[11] The cost per kilowatt at its top capacity was $112. Given inflation, this was analogous to the Oyster Creek cost of $108. GE and Westinghouse executives were concerned that other orders were not being received.

But by the autumn of 1965, "General Electric's gamble that a competitively priced nuclear plant would prove irresistible to many American public utilities . . . proved successful."[12] Utilities across the nation began to place orders for nuclear reactors, even ordering them two at a time. In 1965, three GE BWRs and three Westinghouse PWRs averaging close to 700 MW(e) were ordered.[13] In addition, Gulf-General Atomics received an order for a 330 MW(e) HTGR. "Probably the single most sensational event in the American nuclear breakthrough in 1965-67 was the order placed by the Tennessee Valley Authority in June 1966 for two 1065 MW boiling-water reactors. By far the most remarkable feature in this case was the fact that this plant was to be built and located in Alabama in the heart of a coal-producing area."[14] According to Burn, the price of coal at the Browns Ferry location was estimated at less than 19 cents a million BTUs, but the "TVA had concluded after an exhaustive analysis that the cost of a large nuclear plant would be slightly less than that of an equivalent coal-fired plant, and the cost of power from such a plant would be 2.37 mills per KWH, while the power from a coal-fired station would cost 2.83 mills."[15] In addition to the two Browns Ferry plants, another eighteen LWRs were ordered in 1966, for a total of twenty-six in a year and a half. Twelve of these plants were purchased on a fixed-price, turnkey basis.[16] After

---

[11] Burn, *The Political Economy of Nuclear Energy*, p. 26.

[12] Jensen, *Nuclear Power*, p. 101.

[13] The decisions that led to the specific orders were (and continued to be) particular to each utility. See Perry et al., *The Light Water Reactor*, Sec. II, and, for the utilities' perspectives, Arturo Gándara, *Utility Decisionmaking and the Nuclear Option* (Santa Monica, Calif.: The Rand Corporation, R-2148-NSF, July 1977).

[14] Jensen, *Nuclear Power*, p. 101.

[15] Burn, *The Political Economy of Nuclear Energy*, p. 30. It should be pointed out that the TVA, being federally owned, borrows money at lower interest rates and pays lower taxes than other utilities. These factors would have made a difference in the Browns Ferry cost estimates.

[16] Perry et al., *The Light Water Reactor*, p. 23, elaborate on the GE and Westinghouse contractual agreements.

these, GE and Westinghouse refused to offer further turnkey contracts. In 1967, another thirty reactor units with an average capacity of over 850 MW(e) were ordered.[17]

Clearly, nuclear energy for the generation of electricity had made its breakthrough. As Hogerton suggested, it has been accepted, even embraced, by the U.S. nuclear and electric utility industries.[18] This acceptance of the PWR was reflected in the 1967 AEC report, which stated that "in view of the striking progress that had been made in the development of light and boiling-water reactors and their construction on a wide commercial scale (i.e., by General Electric and Westinghouse), the Commission would in future concentrate its resources on [developing] fast breeders."[19] For all intents and purposes, the LWR had been developed, demonstrated, and accepted in the United States. The commercialization of the reactor on its intrinsic merits had been accomplished.

The Soviet Union

The point of major decision for the Soviet dissemination of nuclear reactors appears to have been in the late 1960s, when large numbers of both PWRs and LWGRs were opened (see Table 5-1). In 1965, nuclear power stations only accounted for 0.3 percent of the Russian electricity output.[20] Between 1971 and 1975, 3.8 million nuclear-generated kilowatts were built, which was 6.5 percent of the total additions to the Soviet electricity network.[21] By 1973, the percentage of nuclear-generated electricity was reported to have risen to 1.4 percent, with every indication that "in the future an increasing percentage of total electrical power production will be from nuclear reactors."[22] Shabad recently

---

17 Plant contract and MW(e) estimates are from U.S. Atomic Energy Commission, "Central Station Nuclear Plants," *The Nuclear Industry—1974*, Table 1.1, pp. 8-9.

18 Bupp and Derian, *Light Water*, refer to this as the "great bandwagon market."

19 Quoted in Jensen, *Nuclear Power*, p. 103.

20 Ibid., p. 82.

21 Theodore Shabad, "Soviet Union Steps Up Installation of Nuclear Power Plants," *New York Times* (January 17, 1977), p. D5.

22 Pryde and Pryde, "Soviet Nuclear Power," p. 26.

estimated that between 1975 and 1980, "the proposed completions are 13.7 million kilowatts, or 20 percent of added capacity."[23] Pairs of large reactors, rated at over 1,000 MW(e), are planned for sites near Leningrad, Smolensk, Kiev, and Kursk.[24]

Like the United States, the Soviet Union chose to disseminate two reactor types, the PWR and the LWGR. Unfortunately, there is little record of the Soviet decisionmaking rationale or processes that determined this choice. A verisimilar explanation is that both reactors had been shown to be economically and operationally viable and that each had its own particular advantages and disadvantages. For example, the PWR was probably less expensive to build and operated at slightly higher thermal efficiencies; the LWGR was the first Soviet reactor model, was more efficient as a plutonium producer, and might have an advantage in safety requirements. Given that neither reactor presented a dominant solution and that both had probably marshalled a sizable group of advocates, one can speculate that Soviet policymakers preferred to let both reactors be disseminated rather than choose one over the other. The BWR option was, however, effectively eliminated.

Two factors in the Soviet nuclear calculus are clear. Economics and the geographical vastness of the Soviet Union played a major part in the decision to build a network of nuclear reactors, especially in the European portion of the Soviet Union. Nearly 90 percent of Russia's fossil fuel resources are located in the Asian part of the nation,[25] and Emelyanov declares that "the necessity for building those [nuclear] stations follows above all from the

---

[23] Shabad, "Soviet Union Steps Up Installation of Nuclear Power Plants," pp. D1, D5. If present goals were met, this would give the Soviet Union 19.4 million kilowatts of nuclear-generated electricity in 1980. To place this in perspective, this would be approximately 25 percent of the projected U.S. total and "equivalent to capacities scheduled to be in service in France or in West Germany by 1980."

[24] See Gillette, "Nuclear Power in the U.S.S.R.," p. 1003; also Petrosyants, "The Atom on the Scales of Progress," p. 21; for the most recent estimates, see ibid., and Ann MacLachlan, "The Russians are Coming . . . on Strong in Nuclear," *Energy Daily* 6, 200 (October 17, 1978):3-4.

[25] Wilczynski, "Atomic Energy for Peaceful Purposes . . . ," p. 569.

fact that nearly all hydroelectric resources in the Soviet Union's European part are now utilized." Emelyanov continues:

> The main characteristic of atomic power stations is that their location is not in any way tied to their sources of fuel, and atomic power has an enormous thermal equivalent compared to fossil fuel. It is therefore much cheaper to transport than fossil fuel. This is where all other types of fuel are inferior to it.[26]

According to a 1973 Polish survey, the capital cost of a 440 MW(e) PWR such as those at Novovoronezh were 230 rubles per kw, which was about 90 percent greater than an equivalent coal- or oil-fueled generating station. The operating costs are between 0.6 and 0.7 kopecks/KwHr compared with 0.8 and 0.9 for the fossil-fueled station. For a 1,000 MW(e) Novovoronezh-type PWR, the capital costs were projected to decline to 170-200 rubles per kw and the operating costs to 0.45-0.50 kopecks/KwHr.[27] Emelyanov states "that an atomic power station with a unit power of 1,000,000 kilowatts [1000 MW(e)] or more will cost about the same to construct as a modern thermal power station."[28] Clearly, then, perhaps the most important reason that the Soviet policymakers decided to disseminate their power reactors was that they were forecast to be economically superior to fossil-fueled stations, particularly given the long distances from western Russia to coal and oil deposits. In short, they were becoming fairly efficient technologies for producing electricity.

A third reason, as Jensen suggested above, might be national prestige. Certainly the Soviets have actively offered their nuclear reactors on the world market, particularly to the satellite nations; however, India is the only nonsatellite nation to have purchased one.[29] An interesting observation is that the Soviets, to date, have only exported the PWR and reserved the LWGR for

---

[26] Emelyanov, "Nuclear Energy in the Soviet Union," p. 41. Petrosyants, "The Atom on the Scales of Progress," p. 19, seconds this contention: "Although the world as a whole still has plenty of coal, there is the cost of transportation to be considered." Also see Knorre, "Frontiers of Technology: U.S.S.R.," p. 317.

[27] Cited in Wilczynski, "Atomic Energy for Peaceful Purposes . . . ," p. 572.

[28] Emelyanov, "Nuclear Energy in the Soviet Union," p. 41.

[29] See Duffy, "Soviet Nuclear Exports."

domestic use. Although it is impossible to be certain why, one might speculate that the LWGR has not been exported because of its efficiency as a plutonium producer. A second possible reason is that the LWGR is much larger and harder to prefabricate than the PWR, thereby making it more difficult to construct at sites far removed from the supporting heavy industry. The comparative cost and size advantage of the PWR might be another reason, but it is not sufficient to exclude the LWGR.

Beginning in the late 1960s and continuing into the 1970s, the Soviet Union carried out a program of the widespread dissemination of nuclear reactors. The specific reasons and their supporting evidence why are less important than the recognition that the Soviet decisionmakers did indeed choose to accept and disseminate the end product of Russia's nuclear reactor RD&D processes.

Canada

In Canada, Ontario Hydro's purchase of two 514 MW(e) CANDU reactors for Pickering in 1964 was a forerunner of the general acceptance of that reactor. This was substantiated in 1967 when Pickering units 3 and 4 were ordered. Ontario Hydro agreed to "finance the entire $256 million cost of the additional stations."[30] The initial capital cost of the four Pickering generating stations was $746 million, which, according to Canadian spokesmen, was a "large financial risk . . . taken in order to maintain the momentum of Canada's nuclear program."[31] Unit 1 achieved full power three months after becoming critical; unit 2 reduced this figure to fifty-three days, and units 3 and 4 cut the margin to eighteen and

---

30 "Ontario Hydro's Nuclear Activities," p. 502.

31 L.W. Woodhead and L.J. Ingolfsrud, "Performance of Canadian Commercial Nuclear Units and Heavy Water Plants," *Annals of Nuclear Energy* (London: Pergamon Press, 1975), p. 767. Ontario Hydro had the same cost estimation and inflation problems that plagued the other national programs. The 1968 estimated costs (ibid.) for the four Pickering generating stations were $528 million, $218 million less than the 1975 reported price, an increase of over 40 percent. The U.S. experience is the best documented; see Perry et al., *The Light Water Reactor*, especially Sec. IV.

twelve days, respectively.[32] Woodhead and Ingolfsrud claim that "This is a dramatic indication of the soundness of the basic design and the total design, construction, and commissioning effort. It also demonstrates one of the many advantages of building multiple identical units at one location."[33] An additional vote of confidence was provided when Hydro Quebec ordered a moderate-size, modified CANDU reactor. Gentilly 1 was rated at 250 MW(e) and used boiling water as a coolant instead of heavy water.

Although the absolute number of reactors ordered was not great, by the early 1970s Canada had clearly chosen to base a large percentage of its electricity generating capacity on the Canadian-designed CANDU reactor. The magnitude of this commitment is demonstrated in the statement of J.L. Gray, President of AECL:

> On a per capita basis, Canada leads the world in nuclear power generating. For December 1973, this was some 45 percent more than in Britian, 85 percent more than in the U.S. One of the world's largest nuclear stations, at Pickering in the Province of Ontario, together with two smaller units, provides nearly 20 percent of the electricity consumed in the province.[34]

In 1974, nuclear power accounted for approximately 5 percent of all the electricity generated in Canada.[35]

The primary reason for this acceptance appeared to be the economics of nuclear energy with large-scale plants—for example, Pickering—especially the fuel cost differential. Morison costs the Pickering stations at $365 per kilowatt, or "1.9 times the cost of an equivalent sized coal fired station built at the same time."

---

32 G.M. Morison, "CANDU Nuclear Power in Ontario," *Nuclear Engineering International* 19, 217 (June 1974): p. 486.

33 Woodhead and Ingolfsrud, "Performance of Canadian Commercial Nuclear Units . . . ," p. 767. This substantiates an early prediction of J.L. Gray: "A very significant reduction will occur in the capital cost of nuclear plants by repetition of construction of similar units." J.L. Gray, "Frontiers of Technology: Canada," *Nuclear Engineering* 9, 100 (September 1964): p. 316.

34 J.L. Gray, "The CANDU-PHW Programme," *Nuclear Engineering International* 19, 217 (June 1974):476.

35 The Honorable Donald S. Macdonald, "Canada's Role in World Developments of Nuclear Energy," *Nuclear Engineering International* 19, 217 (June 1974):475.

His claim is that the relative fuel costs over the long run make the CANDU more economical.[36] Woodhead and Ingolfsrud give a similar cost comparison (Table 7-1) for the Pickering stations with a comparably sized Ontario Hydro coal-burning station, Lambton.

The figures are based on an 80 percent capacity factor, using low sulphur coal, Ontario Hydro capital depreciation conditions of thirty years' reactor life and 8 percent capital interest rate, and are expressed in 1975 dollars. The fuel costs for Lambton (1) were the 1975 cost of coal in the inventory, and the Lambton (2) price represented the 1975 replacement value of coal, reflecting the rather drastic rise in the price of fossil fuel since 1973. The cost comparisons do not completely reflect the financial and operating situation because a coal station can alter its output as a function of the demand, whereas the nuclear station normally operates at a maximum capacity to supply a base load requirement of the system.[37]

Canadian commentators suggest that their extensive D_2O plant construction program has alleviated their heavy water supply problem.[38] However, Canadian estimates in the mid-1960s were equally optimistic and were proven to be startlingly wrong.[39] Indeed, in 1975, AECL still had to purchase close to 150 tons of heavy water from Sweden and the Soviet Union to meet its domestic and contracted external (e.g., Great Britain) requirements.[40] The most recent estimate is that Canada can meet all its D_2O commitments by the early 1980s.[41]

In brief, Canadian industry[42] and utility chose to accept the results of its national nuclear reactor RD&D and initiated a

---

36 Morison, "CANDU Nuclear Power in Ontario," p. 487.

37 This point is made by McIntyre, "Natural-Uranium Heavy-Water Reactors," p. 19.

38 Haywood, "Heavy Water Production," pp. 506-507.

39 Cf. Holt, "Canadian HWR: Economic Assessment," pp. 461-463, and Law, "Heavy Water Production," pp. 510-512 (June 1966 and 1968, respectively), with Haywood, "Heavy Water Production."

40 Atomic Energy of Canada, Ltd., *Annual Report 1974-75* (Ontario: AECL, 1975), p. 13.

41 J.A.L. Robinson, "The CANDU Reactor System: An Appropriate Technology," *Science* 199, 4329 (February 10, 1978):657-664.

42 James A. Weller, "What Nuclear Power Means to Canadian Industry," *Nuclear Engineering International* 19, 217 (June 1974):500-503. Canadian in-

Table 7-1. CANDU versus Coal: Unit Energy Cost (UEC) Comparison *(mills per kilowatt hour)*

	Pickering	*Lambton (1)*	*Lambton (2)*
Capital UEC	4.60	1.70	1.70
O&M UEC	1.10	0.96	0.96
Fueling UEC	0.98	10.60	13.52
Heavy water upkeep	0.35	–	–
Total UEC (m$/KwHı)	7.03	13.26	16.18

Source: Woodhead and Ingolfsrud, "Performance of Canadian Commercial Nuclear Units . . . ," p. 769.

relatively widespread program of dissemination both at home and in the international market (e.g., India;[43] see Table 5-5), in spite of the possibility of continued D_2O supply problems.[44] *Nuclear Engineering* almost wistfully editorialized on the Canadian scene, "the main home customer has always believed in the system the main research organization is developing—a situation almost unheard of in other parts of the world."[45]

Federal Republic of Germany

The last of the nations whose nuclear actors decided to accept the results of their previous research was the Federal Republic of Germany. In West Germany, the decision was less the acceptance of a specific reactor type than the culmination of the strug-

dustry can hardly be described as parochial; realizing the majority of the world's nuclear power plants require enriched uranium fuel elements, Brinco Ltd. is planning to operate a gaseous diffusion plant to help meet the growing demand. See M.Y. el Baroudi and P.A.T. Keeping, "A Canadian Initiative in the Uranium Enrichment Field," *Nuclear Engineering International* 19, 217 (June 1974):508-509.

43 The Canadian experience has not always met the government's expectation in matters of nuclear proliferation despite its self-imposed safeguards. See Robert Gillette, "India: Into the Nuclear Club on Canada's Shoulders," *Science* 184, 4141 (June 7, 1974):1053-1055. Gillette concludes his article: "From the Canadian point of view, it has begun to look like a case of technological charity gone sour."

44 See Jeff Carruthers, "Canada Finally Clinches Sale of Two Reactors to Romania," *Energy Daily* 6, 202 (October 19, 1978):4.

45 Editorial, "Nu-Can CANDU," *Nuclear Engineering* 13, 145 (June 1968):497.

gle within West German industry over who would dominate. In 1964 German private industry decided to construct a BWR for the Lingen and a PWR for the Obrigheim power stations, and the German government and industry refused to join with the French in building a jointly sponsored gas-graphite reactor at Fessenheim. Further important evidence was provided in the summer of 1967 when German utilities placed commercial orders for two additional LWRs in the 600-700 MW(e) range, a BWR at Wurgassen and a PWR for Stade, with AEG and Siemens, respectively. Unlike earlier German reactors, these had no direct federal subsidies. An official in MAE explained:

> [the] preponderance of light water reactors in Germany is due to several factors. First of all, our private utilities place an over-riding emphasis on costs, and especially on construction costs. Secondly, traditional links between German and U.S. manufacturers made the transfer of the technology relatively easy. Thirdly, bilateral agreements between the U.S. government on one side and both the German government and the Euratom commission on the other secured the supply of fuel.[46]

As the German utilities became willing to purchase power reactors,[47] it became clear that the earlier funding arrangements were not sufficient to maintain an advanced nuclear R&D program. It was "evident that the necessary finance involved was too high compared with the possibilities of the industry and the States."[48] For this reason, the Third Nuclear Program (1967 to 1972), which was announced in December 1967, increased both the absolute amount and percent that the federal government contributed to the German nuclear development program. During Germany's Second Nuclear Program, the *Bund* had budgeted 3,160 million marks for nuclear RD&D. The financial commitment for the Third Nuclear Program was projected to be 4,900 million marks, an increase of over 55 percent. The amount of the *Länder* contribution to the Second Nuclear Program, 1,100 million marks, remained unchanged for the Third Program, an ef-

---

46 Finke, "F.R.G. Research Activities," p. 743.

47 See Mackenthun, "Electricity Supply Industry," p. 748.

48 C. Salander, "Nuclear Research Centres in the Federal Republic of Germany," *Nuclear Engineering International* 18, 291 (February 1973):97.

fective reduction of roughly 11 percent.[49] The federal support of the nuclear research centers rose to almost 90 percent of their annual budgets.[50] The federal support and emphasis appeared to be a reflection of a wider support base than might have existed previously. A British trade journal noted at the time that the Third Program "is born of public opinion, while the earlier programmes have only been something akin to recommendations by the experts."[51] In short, the announcement of the German Third Nuclear Program manifested the centralization of government sponsorship of nuclear RD&D and, concomitantly, more effective government control.

At approximately the same time, AEG and Siemens, as separate members of construction consortia, were competing for the French-Belgian SEMO LWR reactor to be built at Tihange as a follow-on to the SENA reactor at Chooz. In December 1967, the contract was awarded to a Belgian firm and the French Framatome.

The West German government had never assumed an aggressive research and development stance toward atomic research. However, by 1968, "Federal influence was unmistakable in promoting industrial partnerships both at home and with foreign firms."[52] International associations were promoted to deflect potential criticisms of the German nuclear program, especially in light of the feared discrimination by NPT. Domestic consolidation was emphasized for economic reasons. Therefore, when AEG and Siemens saw themselves losing the Tihange bid and dividing the domestic market, consolidation was governmentally urged and quite imminent. Negotiations started in the fall of 1967, and in April 1969, the two companies announced the creation of the giant nuclear engineering subsidiary, *Kernkraftwerk Union* (KWU). Shortly thereafter, Siemens purchased the outstanding shares of Interatom from North American Aviation. Multiple reactor orders for plants both at

---

49 "Better Basis for German Third Programme," *Nuclear Engineering* 13, 142 (March 1968):159.

50 Salander, "Nuclear Research Centres . . . ," p. 98.

51 "Better Basis for German Third Programme," p. 159; also see Schmidt-Küster and Popp, "The German Nuclear Programme."

52 Nau, *National Politics and International Technology*, p.87.

home and abroad followed (see Table 5-2). Siemens allowed its license agreements with Westinghouse to expire in June 1970 and AEG did likewise when its license with GE expired in 1973.[53] Nau summarizes the amalgamation and its implications for the German nuclear industry:

> The consolidation of German industries, culminating in 1969 in the establishment by Siemens and AEG of a joint nuclear subsidiary, KWU, did not reflect exclusively economic exigencies. Industrial consolidation was also a way of expressing German national interests in an increasingly economic, as opposed to strategic, world context.[54]

The transition to the dissemination stage in West Germany was perhaps the least complicated of any of the considered nations because German industry and the government had jointly decided to enter the commercial nuclear field when nuclear R&D was initiated in the mid-1950s. Much of the atomic research was performed by firms that displayed a strong commercial interest in reactor RD&D. In addition, the German nuclear actors began their nuclear research when many of the early uncertainties about nuclear energy had been resolved and the market for energy reactors seemed fairly certain. The research strategy reflected the German late start and the intention to commercialize reactors as soon as possible, both of which meant that German industries would have to avail themselves of existing exogenous technologies.[55] German firms developed their own nuclear research facilities, which worked closely with (and often staffed) government research centers. This research focused domestic efforts on the advanced technology reactors—such as the HTGR, the THTR, and especially the fast breeder reactor—and left the near term research (and commercial) gap to be filled by technology transfers.[56] Largely because of the domestic industrial competition, the German nuclear industry produced two types of nuclear reactors, chosen because of their advanced state

---

53 Because of licensing agreements, the two reactor divisions of the parent companies could not be completely merged until 1973.

54 Nau, *National Politics and International Technology*.

55 O. Keck, "West German Science Policy Since the Early 1960's: Trends and Objectives," *Research Policy* 5, 2 (April 1976):127-128.

56 See Jensen, *Nuclear Power*, pp. 129-131.

of development, the relative ease of construction (which was particularly important given the German view toward exploiting the export market), and the economies of construction and operation that were becoming evident in the late 1960s.

Reactor Rejections

Great Britain

The dissemination period in Great Britain was defined by the rejection of the British AGR on economic and technical grounds. As discussed earlier, by 1972, over 150 million pounds had been spent on the development of the AGR, not even counting plant construction. Early optimistic British export hopes had been rudely dashed; only two reactors, both of them Magnox, had been sold abroad, and none had been exported for over a decade. In addition, the British nuclear industry, which had been reduced to the duopoly of BNDC and TNPC in 1969, was again in serious trouble. Their work on the AGR had been discouraging, both were apparently lacking in project management skills, and as a result, they were losing money. The fuel shortages of 1973 accelerated the nuclear power malaise, and by late 1973 the CEGB again began to advocate that England adopt the U.S. PWR as its principal power reactor. All of this once again forced the government to confront the issues of Britain's nuclear power industry and the English reactor system.

In the face of rapidly rising energy demands, the government considered a number of alternative reactor designs, including the American LWR, the British steam-generating heavy water reactor (SGHWR), the Canadian CANDU, a HTGR such as the British had been developing with the ENEA-sponsored DRAGON reactor, and the fast breeder reactor.[57] Some argued that England should continue with the Magnox and advanced gas reactors until the FBR was operational, but the FBR target

[57] See Aurum, "Britain Chooses a Reactor," *Bulletin of the Atomic Scientists* 28, 2 (February 1972):45; and "Will CANDU Do for the UK?" *New Scientist* (December 1973), pp. 702-703.

date was increasingly distant and the short supply of British plutonium heightened the uncertainty surrounding the FBR. Technical troubles and cost increases continued to obstruct the AGR's development (see Table 7-2).

In early 1973, a new consortium was announced, the National Nuclear Corporation (NNC). In the report of the 1967 Select Committee of Science and Technology, the sole consortium recommendation had been explicitly rejected because it was thought that the creation of a monopoly industry "would leave the industry without sufficient competitive stimuli and would tend to concentrate power in the hands of the A.E.A."[58] But within five years, the government was forced by economic conditions to recant and created the NNC, which subsumed BNDC, TNPG, and the reactor design teams from the AEA Risley facility. British General Electric Company (GEC), the Westinghouse license in Great Britain, owned approximately 50 percent. AEA held another 15 percent with the proviso that "The Government would have special rights in certain matters where the public interest is clearly involved. These would include the formation of international links and securing that an open purchasing policy is pursued. . . ."[59] The remaining 15 percent was to be privately

Table 7-2. U.K. AGR Cost Escalation
(millions of £)

	Dungeness B	*Hinkley Pt. B*	*Hartlepool*	*Heysham*
Date of original estimate	31 Mar. 65	31 Mar. 66	31 Oct. 68	30 Nov. 70
Original estimate price	89	95	92	142
1976 estimated price	280	140	220	240
Original estimation adjusted for inflation	186	126	158	174
Percent increase (adjusted)	109.0	32.8	71.8	22.5

Source: Simon Rippon, "The AGRs Make a Start," *Nuclear Engineering International* 21, 242/243 (April/May 1976):56.

---

[58] Jensen, *Nuclear Power*, p. 112.

[59] Quoted in "GEC Wins British Nuclear Industry," *Nuclear Engineering International* 18, 203 (April 1973):276.

owned, which meant the existing shareholders of TNPG and BNDC. As a British observer commented, the fact that so many private companies "who have suffered in the past from the vagaries of the British nuclear power policy still are interested even in a small share of a minority stake in the new company reflects a great deal of faith in the long term prospects of nuclear energy."[60]

In late 1973, CEGB published a study contending that the PWR was the cheapest reactor to build and operate and that only the PWR would permit an early and rapid deployment of new and necessary nuclear plants. Not surprisingly, the CEGB study was strongly supported by the NNC, whose principal stockholder held the British license for the Westinghouse PWR.

The second major reactor under consideration was the AGR, but technical design problems continued to plague the system. Perhaps the central problem was the choice of CO_2 as the coolant, which was not as benign and nonreactive at higher temperatures as initially thought. At one point, British nuclear engineers seriously considered as radical a substitution as replacing the CO_2 with helium.

The third major reactor system under consideration was the SGHWR. A SGHWR prototype had been authorized at Winfrith in early 1963 and had been finished in 1967; the reactor reached full power in 1968.[61] The Winfrith prototype SGHWR used heavy water as a moderator, boiling light water as a coolant, and individual Zircalloy pressure tubes in a pressure vessel. The fuel was low enrichment (2 percent) uranium oxide. The reactor was thus a hybrid of the American BWR and the Canadian CANDU systems. Its main drawbacks were the question of whether the 100 MW(e) prototype could be readily expanded to a full commercial-sized reactor within reasonable time and money constraints and its lower thermal efficiencies than the water cooled systems, roughly 40 percent rather than near 30 percent.

---

60 Simon Rippon, "Primary Interest—To See a Healthy Nuclear Industry," *Nuclear Engineering International* 18, 204 (May 1973):383.

61 For early reports on the SGHWR prototypes, see "SGHWR: Advanced, Economic, and Simple," *Nuclear Engineering* 11, 121 (June 1966):468-470. Also see H. Cartwright, "SGHWR: A Darkhorse Competitor," *Nucleonics* 24, 9 (September 1966):60-63.

In July 1974, the Labour government announced its selection of the SGHWR over the PWR. The United Kingdom correspondent for *Science* reported the decision:

> The British proclivity for compromise shines through the Labour Government's recent choice of a reactor for its next order of nuclear-power stations. After considerable delay and controversy the government picked the homegrown SGHWR over an American design originally preferred by Britain's biggest electricity-generating authority and by the public-private corporation that will build the reactors.[62]

The government White Paper also announced a 4,000 MW(e) commercial program of nuclear plant construction during the 1975-1979 time frame. The first plant was to be approximately 650 MW(e) and was scheduled for completion before 1980.

The stated reasons for the SGHWR selection were "reliability" and that the "SGHWR offers particular scope for British nuclear technology."[63] A third unmentioned reason but one generally believed to be pivotal was the growing concern for the safety features of the PWR and the questionable adequacy of its existing safeguards. A fourth point was that the British apparently preferred to cooperate with the Canadians rather than the United States; for example, Canada promised to supply D_2O for the English reactor. A fifth reason was that British labor unions, fearing the loss of jobs to American technology, lobbied strongly within the Labour party for the SGHWR. But perhaps the most important reason was that Britain was unwilling to surrender its noble heritage of nuclear research. Rather than "sell short on the nuclear technology in which all British governments have invested heavily since World War II,"[64] Labour chose to abandon the gas reactor that had so auspiciously introduced the English to the era of nuclear-generated electricity and advance an untested but completely British design.

The dissemination stage was as traumatic for the British as the earlier demonstration period. A second reactor system had pro-

---

[62] John Walsh, "British Choose Own Reactor for Nuclear Power Program," *Science* 185, 4150 (August 9, 1974):511.

[63] Quoted ibid.

[64] Ibid.

ven to be inadequate and had left the English once again to select a new reactor system to bridge the projected power deficiency until the FBR were operational. Rather than select the proven American reactor system, which was also the preference of the nuclear and electric utility industries, the government had chosen a system tested and developed in the United Kingdom, thereby exposing itself to persistent charges of "nuclear nationalism." The end result of the British inability to produce an economically and technically viable reactor was that Britain still did not have a proven and disseminated reactor by 1975, close to thirty years after British authorities made the decision that the United Kingdom should reap the peacetime benefits of nuclear technology. These failures almost certainly contributed to the 1976 recommendation of a royal commission that the further expansion of the British nuclear power plant system "be postponed as long as possible"[65] and the renascent interest in the PWR expressed by the chancellor of the exchequer.[66]

France

France in the late 1960s faced a situation similar to England and very analogous to the British dissemination experience just described. In late 1967, the deGaulle government had reaffirmed the French allegiance to the French gas-graphite reactor, but there were strong indications (e.g., the two Franco-Belgium reactors) that the government was backing away from that system and toward the acceptance of foreign nuclear technology, in this case, the American PWR. The years 1968-1969 marked the critical period when the French nuclear actors were forced to

---

65 Sixth Report of the Royal Commission on Environment Pollution, *Nuclear Power and the Environment*, London, 1976. The quotation is from Peter T. Kilborn, "British Commission Urges a Delay in Widening Use of Nuclear Energy," *New York Times* (September 23, 1976), pp. 1, 5; an excerpt is found on p. 4. Also see Nicholas Wade, "European Doubts About Nuclear Power," *Science* 194, 4261 (October 8, 1976): p. 164.

66 Nigel Hawkes, "Britain Opting for U.S.-Style Reactors—Maybe," *Science* 199, 4330 (February 17, 1978):755-756.

decide if their domestically developed reactor system would be disseminated or rejected in favor of the LWR.

In 1968, it was apparent that the French nuclear community was seriously split over the merits of the two reactors. The government, through CEA, continued to champion the gas-graphite reactor; EdF and the nuclear industry, chafing at government-imposed restrictions, favored the LWR. In April, this debate once again reached the public forum when a PEON Commission Report "called, *inter alia*, for the construction by EdF of at least one enriched uranium-type power station of American design, the construction of a heavy water reactor . . . and the construction of an enriched plant."[67] Specifically, the PEON committee recommended that after the planned Fassenheim 1 and 2 nuclear facilities, no new gas-graphite reactors contracts be let until after 1970; that construction begin on a French LWR with a rating power of at least 600 MW(e); and that research on a heavy water, natural uranium reactor continue but in cooperation with other nations.[68] The LWR recommendation was predicated on the availability of enriched U-235 from the United States through its agreements with EURATOM, although the long-term preference was the French Pierrelatte plant or a European facility.[69] Finally, it called for the French nuclear industry and government to move aggressively into the export market, arguing that exports were essential if the domestic industry was to produce nuclear facilities at costs competitive with U.S. and German firms, and especially because the gas-graphite reactor was inherently at a disadvantage because of its large size and initial capital costs.[70] No major changes occurred in the French nuclear context in 1968.

Perhaps in reaction, the PEON committee the following May was even more decisive in its subsequent report. It recommended that all further commitments to gas-graphite facilities be suspended until the Saint Laurent-1 facility could provide operational evidence upon which to evaluate the possibility of addi-

---

[67] Jensen, *Nuclear Power*, p. 128.

[68] *PEON Reports*, April 1968, Tome I, pp. 69-70.

[69] "Securité d'Approvisionement et Independence," Annexe VI, ibid., pp. 88-90.

[70] "Problèmes d'Exportation," Annexe VIII, ibid., pp. 97-98.

tional gas-graphite power reactors.[71] After briefly reviewing other national programs and industrial participants,[72] the committee recommended that additional research be done on a heavy water reactor,[73] but that plans should be immediately initiated to construct four or five light water type reactors, ranging in size from 700 to 900 MW(e), between 1970 and the end of 1975.[74] The reasons for the committee's choice were almost entirely economic. The Saint Laurent reactor design was considered to be so exemplary that it could not be improved upon for

> the veracity of the estimation of construction costs and of the other components of the present value financial evaluation. [Therefore,] it became obviously unrealistic to anticipate any substantial cost reduction for other reactors of this type, also considering the inherent limitation of the output per reactor, which makes it practically impossible to build them over 600-700 MW(e). At the same time the construction and fuel fabrication costs were rapidly decreasing for the water reactors, and their proposed unit size was increasing up to the 1200 MW(e) range, which is quite suitable for the EdF grid size. At the same time also the fuel-oil prices and availability were changing, so that a policy of mass production through nuclear power was appearing more desirable, and there was no time to gain experience with prototypes.[75]

The PEON report also suggested that the French nuclear industry be restructured to one reactor design and construction group that would emphasize one reactor type, preferably the LWR. This was seen as necessary to give the company a sufficient number of orders to survive. The implication was that France could not afford to pursue two different conventional reactor technologies simultaneously.[76]

In November 1969, the Pompidou government announced that France would no longer develop or build gas-graphite reactors.

---

[71] *PEON Reports*, May 1969, Tome I, p. 103.

[72] Ibid., pp. 116-124.

[73] The emphasis here would be on the CANDU technology; ibid., "Etude d'Une Centrale Nucléaire à Eau Lourde," Annexe 3, pp. 125-127. Also see Editorial, "Couture Cures Confuse," *Nuclear Engineering International* 14, 153 (February 1969):72.

[74] *PEON Reports*, May 1969, Tome I, p. 137.

[75] J. Stolz, "Nuclear Future for Electricité de France," *Nuclear Engineering International* 17, 196 (September 1972):960.

[76] *PEON Reports*, May 1969, Tome I, p. 137.

"Pompidou ordered the immediate preparation of tenders for a 600 MW(e) light-water reactor and indicated that the Sixth Modernization Plan scheduled to begin in 1971 would call for the construction of only three reactors, the types to be determined by the progress of the first light-water plant and French studies on the heavy-water reactor family."[77] He also established a committee to investigate the changing responsibilities of CEA in the context of the French nuclear programs as the French military atomic R&D was phased down and Pierrelatte was completed. His decision was obviously disputatious, especially within the CEA bureaucracy, which by this time had grown to a staff of over 30,000 people and a budget of close to five billion francs (or about 3.5 percent of the French government budget).[78] The reasons for Pompidou's decision were mostly economic, but they also reflected the technical superiority of the American design. The LWR was proving to be less expensive in both capital and operating costs than the gas-graphite types.[79] Second, French industrial spokesmen argued that the PWR would be more exportable than the gas-graphite models. Third, EdF contended that the worldwide acceptance and the U.S. and West German commitment to the LWR guaranteed that France would have less technical trouble building a LWR system than could be expected from any other reactor. Finally, both Prime Minister Jacques Chaban-Delmas and CEA High Commissioner Francis Perrin stated that the relative economics of the gas-graphite reactor had been undermined by the "spectacular" drop in oil prices, almost 40 percent in the previous five years. Perrin predicted that future French nuclear *filieres*, which would depend on U.S. or U.S.-German licenses, "would prove lower than conventional fuel-oil stations."[80]

Having made the major decision, the French government then

---

[77] Nau, *National Politics and International Technology*, pp. 84-85.

[78] Commissariat à l'Energie Atomique, *Rapport Annuel 1970* (Paris: CEA, 1971), p. 132.

[79] Still, French sources were careful to stipulate: "it must be clearly understood that the gas cooled reactors are technically satisfactory, even if they cannot compete with the water reactors on economic grounds, at least not in the French economic conditions." Stolz, "Nuclear Future for Electricité de France," p. 691.

[80] Both are quoted in "Committee Probes CEA's Future as EdF Gets Enrich-

turned to the reorganization of the nuclear industry, using the bidding for the first LWR as leverage. Framatome and SOGERCA (composed primarily of CGE, Alsthom, and GAA) emerged as the new French nuclear consortia, with Framatome receiving the contract for Fessenheim-1, a 900 MW(e) PWR, after a complicated set of negotiations with Westinghouse, and SOGERCA obtaining a GE BWR license.[81] In 1971, Framatome received contracts for an additional three reactors, Fessenheim-2, a 900 MW(e) PWR, and Bugey-2 and -3, each a 925 MW(e) PWR. Still, it became apparent that Framatome was not able to manage the construction of the French PWRs without American expertise. Therefore, in 1972, the French government permitted Westinghouse to buy into the French nuclear industry, and the *Compagnie Nucléaire Francaise* (CNF) was formed with Westinghouse owning 45 percent, Framatome holding 51 percent, and a Belgian company the remainder. As Walsh has commented, "The ultimate decision by the French government on nuclear industry can be seen as a victory for pragmatism over economic chauvinism."[82] SOGERCA was given contracts for two BWRs and several options by EdF in 1973, which were to be built under a GE license, but financial difficulties led the French to cancel all plans in the near future for the BWR, which reduced the SOGERCA members to a position of being suppliers.

The French dissemination stage saw France "set about to free its civilian program from constraints imposed at the start of the decade by military consideration."[83] The decision was made not to disseminate the domestically designed gas-graphite reactor. However, unlike the British when England abandoned the Magnox and advanced gas reactors, France chose to utilize American technology and wholeheartedly embraced the PWR. Given that France had long been the leading proponent of European and especially French technologies to the complete exclusion of U.S. nuclear technology, this was a startling switch, par-

---

ment Go-Ahead," *Nuclear Engineering International* 15, 165 (February 1970):76.

81 See F. Torrisi, "Recent Structural Reorganization in the French Nuclear Industry," *Nuclear Engineering International* 17, 196 (September 1972):702-704.

82 Walsh, "Nuclear Power: France Forges Ahead . . . ," p. 340.

83 Nau, *National Politics and International Technology*, p. 85.

ticularly because the LWR would be partly dependent in the short run on American enriched uranium.[84] The explanation was offered that the reliance on the American LWR would only be an interim measure until the French FBRs were in operation. Still, France chose to turn its back on its own reactor technology. A French observer commented at the time:

> There is some irony in the fact that, by deciding to develop light water reactors, France, whose (nuclear) program was strongly influenced by the desire for national independence, caused its industrial groups to be more tightly linked to their U.S. licensers than any other European manufacturers, such as Siemens. . . .[85]

A comparison between the French and British decisions is inevitable. Both nations faced the same situation. The reactors each had developed were proving to be technically outmoded and increasingly expensive, the coveted exports market was disappearing, and neither nation's private nuclear construction industry was able to remain competitive. England had long boasted of its traditional and "special relationship" with the United States but chose to continue its nuclear program with further development of its own reactor systems in the face of a clear and stated preference for the U.S. LWRs by the electric utility and the nuclear construction industries. France had largely developed its nuclear capabilities motivated by national pride and pointedly emphasized its political distance and technological independence from the United States, but it chose to abandon almost twenty-five years of French slow fission research and accept the American design. In England, AEA's preference was decisive; in France, CEA's preference fell before the arguments of the nationalized electric utility and the government-sponsored nuclear industry. In short, given similar circumstances, the British and the French came to diametrically opposed resolutions as different coalitions of nuclear participants proved successful.

---

[84] "Committee Probes CEA's Future . . . ," p. 77.

[85] Quoted in Bupp and Derian, *Light Water*, p. 69.

Summary

The dissemination period represented in almost every case a critical selection point at which each nation decided to accept or reject a given reactor type. Although each nation's choice was motivated by the previous RD&D, critical events and perceptions during this final period were directly responsible for the decision to accept or reject a given reactor.

In the United States, the growing evidence of the economics of nuclear power generation were underlined by the Oyster Creek contract, the inability of the fossil-fuel industry to lower its prices further, and the exigencies of the environmental movement. However, even with GE and Westinghouse's proven willingness to offer extremely attractive contracts and with other firms entering the market, the LWR owed its almost sudden acclamation in 1966-1967 to what amounted to a bandwagon effect within the utility industry. Once a few American electricity utilities began to purchase nuclear generating plants, orders were placed at such a rate that the reactor industry could not keep pace. With little change in the political or economic conditions, previously reluctant utilities began ordering reactors for seemingly no additional reason except fear of being left behind.[86] This continued in the face of rapidly rising construction costs, expensive safety regulations, and a growing number of environmental lawsuits.

In the Soviet Union, the standardization of reactor designs on the PWR and the LWGR and their increasing size resulted in a general lowering of the cost of nuclear-generated electricity. This was particularly important for the western regions of the Soviet Union, which were far removed from alternative energy sources. The accelerating pace of the Soviet building program in the late 1960s and early 1970s suggests that the dissemination of power reactors in the Soviet Union was very much a reality.

In retrospect, the Canadian dissemination period did not seem to be as critical as in the United States and the Soviet Union. Ontario Hydro's ordering of Pickering generating stations 3 and 4

---

86 On this perception, see Gándara, *Utility Decisionmaking and the Nuclear Option.*

and Hydro Quebec's contract for Gentilly-1 demonstrated that the CANDU system had been accepted in Canada, a selection later verified in the international reactor market by purchases in Korea and Latin America.

The West German dissemination period was, like Canada's, an uneventful continuation of earlier reactor RD&D. However, it differed from the other nations' dissemination periods in two important aspects. First, the German dissemination of a nuclear reactor was not based on domestic technology; the preferred German LWR was originally designed in the United States and was being built with licenses from U.S. firms. Second, the dissemination was marked by an increased involvement by the West German government whereas the American and Canadian commercializations were characterized by a general withdrawal of the government sector from the private sector.

The British dissemination stage was unfortunate in almost every aspect for most of the English nuclear actors. The earlier U.K. decision to develop the AGR was revealed to have been ill-advised as technical difficulties and cost escalations undercut AEA's 1968 appraisal. The British nuclear industries—which had undergone two severe retrenchments—and the nationalized utility both openly lobbied AEA to abandon the AGR and adopt the American light water technology, but for reasons more of atomic patriotism than of financial or technological efficiency, AEA once again chose to develop another reactor model that had operated only as a prototype. The dismal experience of extrapolating the AGR prototype to the full-scale commercial reactor was apparently overlooked in favor of the home-bred SGHWR development program.

Finally, the French situation at the beginning of the dissemination period was analogous to that of the British. The gas-graphite reactor, the product of France's nuclear RD&D program, was seen by the French nationalized utility and industry to be significantly more costly than the competing U.S. and West German light water technologies. Although CEA was adamant in excluding foreign nuclear technology—with the partial exception of Canadian HWR knowledge—the force of EdF and industry's arguments about the domestic costs and the potential loss of the export market were decisive. Thus, France rejected its domestically developed reactor and accepted a foreign technology, even at the risk of a short-term dependence on foreign sources of fuel, a condition earlier governments had explicitly rejected. Perhaps the most interesting argument raised was that the LWR technology would be the easiest to assimilate because of the previous work done by American and German industries. This, of course, closely parallels the strategy used when West German industry was attempting to move into the reactor market as expeditiously as possible.

Part III: Conclusions

8

A Review of the Analytic Framework and Policy Observations

This chapter serves three purposes. First, it considers the levels of analysis presented in Chapter 1 to determine the most useful analytic level for understanding and structuring the development and dissemination of new, large-scale civilian technologies. Second, it synthesizes the data from the six national nuclear power reactor development programs in the context of the analytic framework to extract potential policy implications. Special emphases are placed on the discussion of multiple national objectives, their possible ordering, and how they affect a nation's development strategies and progress. A third task evolves from these first two, the evaluation of the analytic construct as both "intellectual" and policy tools. In other words, does the framework or its possible revisions have any value as a heuristic device or an aid to policymakers for nuclear technology and other technology as well?

Levels of Analysis

Chapter 1 presented three possible levels or models of analysis for examining a national technology development: the national actor model; the organizational model; and the individual model, with particular emphasis on the individual's cognitive limitations and role within a bureaucracy. This analysis selected the organization or institution and their interactions as the most useful perspective, arguing that the national actor level conceals too much pertinent information while the individual level requires so much information that it would overwhelm a study of this scope, even assuming that the necessary information was obtainable. A review of the six national development experiences confirms the choice of analytic levels. The national actor model would have been hard pressed, perhaps fundamentally disfigured, to have usefully explained the variant development patterns observed. The national, unitary decisionmaking state model might have elucidated the sudden French reversal of national position when the French government abandoned the indigenously developed gas-graphite reactor and adopted the long-resisted American LWR as a rational, change in government policy dictated by economics. However, the national actor model would have had great difficulty explaining why the British, faced with a similar situation, chose the contrary alternative. Likewise, it would have been unable to rationalize why the Soviet Union and the United States "illogically" continued the advanced development of multiple reactor models well into the demonstration period. Also, of course, the national actor model and its implicit portrayal of a nation's decisionmaking processes as a "black box" is not relevant to the policymaker hoping to gain some understanding of how policy was and should be made. To be useful for policy, a theory must have more than a predictive element; it must predict at a level of detail or generality offering policymakers real choices that may be implemented with available policy levers. Finally, a policy theory should explicitly recognize the normative elements underlying policy choices.[1] These requirements are not readily obtainable in the national actor level of analysis.

---

[1] Milton Friedman, "The Methodology of Positive Economics," in his *Essays in Positive Economics* (Chicago: University of Chicago Press, 1963), argues that predictability is the only valid test of a model. A broader set of criteria is suggested by Tjalling Koopmanns, "The Construction of Economic Knowledge," in

At the other end of the spectrum, the individual level of analysis clearly would have overwhelmed the analytic framework. More to the point, it is not clear if the resulting details are even necessary for obtaining policy recommendations. In developments of the magnitude examined above, it is uncertain that solitary individuals had more than a transitory influence. Few would question that Admiral Rickover was extraordinarily influential in shaping the early U.S. nuclear reactor development because of his dual positions in the AEC and U.S. Navy reactor programs,[2] but it would be extreme to credit him or any one individual as being indispensable considering the number of pivotal American participants (in both the public and private sectors) and the scale of the development. Certainly few major changes took place in national programs as individuals left or new persons were recruited. Important patterns of development can be discerned by examining individuals in their aggregate levels (i.e., organizations), patterns that are of great policy interest and at a useful level for policy formulation and manipulation. This is not to say that individuals did not make a noticeable difference in some cases; however, for this discussion and especially for deriving policy insights, this level of analysis is inappropriate.

For these reasons, the appropriate level of analysis for this particular inquiry into several national programs is the organization, especially the tension among organizations. A single institution was rarely able to dictate the course of the entire development and dissemination processes. In all cases, the development patterns can be described by the tugging and pulling of the various participating organizations, each manifesting different (at times conflicting) objectives and all influencing the course of the development. Even in the most organizationally monolithic development, that of the Soviet Union, there is evidence of divergent institutional objectives between the atomic researchers

---

Three Essays on the State of Economic Science (New York: McGraw-Hill, 1957), pp. 129-151.

[2] Hewlett and Duncan, *Nuclear Navy*, describe Rickover's role in the development of the pressurized water reactor. Still, they point out that Rickover's influence was largely carried out by the organization he headed; and, of course, a number of different reactor types were developed in the United States.

and heavy industry affecting the Soviet development. The choice of analytic levels of attention is perhaps more critical for the policy researcher than for his or her academic counterpart, for it determines not only the explanations proferred but also the very questions asked,[3] which, in turn, profoundly affect the policy issues raised and policy recommendations rendered.

The organizational perspective facilitates the comparison of the dissimilar development patterns observed among the nations in the sample. From the beginning of the French nuclear reactor development program, CEA dominated the development and demonstration programs, while the user, EdF, was reduced to a role of accepting whatever reactor type CEA dictated. Having invested considerable research in the domestic gas-graphite reactor, CEA was understandably unreceptive to other reactor types, especially the American-developed LWR. CEA was supported for international political reasons by the French government, which resisted the incursion of American political, financial, and technological influence in Europe and France in particular. But by 1968, EdF and the French nuclear industry were able to marshal sufficient evidence to convince the councils of government that the gas-graphite reactor, however Gallic its origins, was decidedly inferior to the less expensive, larger, and more exportable American LWR. The seemingly sudden reversal in French reactor development policy offered by the national actor model is easily understood in terms of interorganizational conflict. That is, beginning in the mid-1960s and culminating in 1968, EdF and its preferred reactor assumed hegemony over CEA and its preferred reactor.[4]

In much the same way, the stubborn British insistence in continuing to develop domestic reactors can be understood by an examination of the relevant organizations. In England, the nuclear development agency, the U.K. AEA, was able to retain its institutional primacy over the vendor industry and the nationalized electric utility (CEGB) in spite of accumulating technical and financial evidence that the AEA-developed family of reactors was

[3] This is discussed by Allison, *Essence of Decision*, ch. 7.

[4] See Bupp and Duncan, *Light Water*, Part I.

less efficient and more costly than the U.S. reactor counterparts.[5] It chose to continue its own research programs rather than import a reactor that not only would have maligned its own developmental capabilities but might also have resulted in a loss of bureaucratic influence and power. In supporting the continued reliance on the Magnox and later AGR, AEA could well have viewed the struggle against the American invader as threatening to its bureaucratic life and certainly its (perhaps even England's) prestige. Whereas in France EdF with its preference for the LWR was able to overcome CEA's insistence on the French gas-graphite reactor (as a last resort, CEA even argued that the French and British should combine reactor development forces to combat the American reactor systems), the British CEGB was unable to effect a similar decision in the British government against AEA.

Just as important as explaining the seemingly "irrational" cases, the organizational model also can help to illuminate the development paterns for programs in which there seemed to be a certain continuity, such as were observed in the United States, the Soviet Union, the Federal Republic of Germany, and Canada. In these nations, the straightforward continuation of the RD&D processes through the dissemination decision can be largely understood by recognizing the cooperation of the important institutional actors throughout the entire process, especially during the transition periods between the phases. In none of these countries was there a debilitating interorganizational struggle between the institutions primarily responsible for the development such as prevailed in France and England in the late 1960s. Nor were there examples in these nations of the long-term dominance throughout the entire process by one agency, such as was seen in the United Kingdom with AEA.

In the United States, West Germany, and Canada, the technology delivery system (TDS) hypothesized in Chapter 2 seemed to be effectively integrated and efficient in delivering an economically and technically viable reactor system without the major organizational difficulties encountered in France and

[5] See Wonder, "Decision-making and Reorganization of the British Nuclear Power Industry."

England where the TDS actors appeared to be much more distinct and disjointed. For the Soviet Union, there is some indication of a minor conflict between the nuclear research community and heavy industry, but by and large, the Soviet counterpart of the Western TDS seems to have worked efficiently. Of course, this judgment is somewhat obscured by the lack of substantive information about Soviet decisionmaking and by the Soviet political centralization, which could have glossed over whatever divergencies occurred in the Russian reactor TDS.[6]

Multiple Objectives and the Developmental Process

The argument for viewing a national program in terms of multiple objectives was presented in Chapter 2. Its validity has largely been substantiated by the national nuclear reactor development experiences related above. The concept helps explain the apparent shifts in the direction of a program. The presence of the multiple objectives reflects the organizational level of analysis; different institutions would be expected to bring their particular objectives into play. For example, for the TDS actors (defined in Chapter 2), the developer or government R&D actor could be viewed as the primary sponsor of the national prestige, science for science's sake, and political equity objectives; the vendor may be perceived as principally emphasizing the development of an efficient technology and its subsequent exportability; and the consumer might have primarily been concerned with obtaining an efficient technology. Thus, the objectives that defined the national reactor development programs can be described in terms analogous to the institutions that shaped them. The multiple objectives, their changing relative positions of importance, and the

[6] That there are often difficulties within the Soviet economy concerning the transition processes between members of the TDS is demonstrated (although not in those terms) by Gertrude Schroeder, "The Economic Reform as a Spur to Technological Progress in the USSR," *Jahrbuch der Wirtschaft Osteuropas* (Munich: Gunter Alzos Verlag, 1971), vol. 2, pp. 343-374; or Joseph S. Berliner, *The Innovative Decision in Soviet Industry* (Cambridge, Mass.: The MIT Press, 1976), especially Part III, "Incentive and Decision Rules." For evidence that Soviet national politics have intervened in technology developments, see Vladimirov, *The Soviet Space Bluff.*

competition among their sponsoring institutions explain the separate national patterns of development dissemination.

During the course of the reactor development and dissemination processes, changes in the availability of resources and requirements dictated changes in institutional and national priorities. In Chapter 2, these altered conditions are aggregated into the development, the demonstration, and the dissemination periods. In light of the national reactor experiences, these three stages can be roughly characterized by the primacy of the developer, the vendor, and the consumer, respectively. In turn, these correspond closely to the roles we have seen assumed by the R&D institutions of the government, the nuclear construction industry, and the electric utilities, respectively. The development period for nuclear reactors was generally initiated and sustained by heavy government involvement in terms of R&D funding and building test reactors; vendor participation in this period was limited to a supplier role, and the utilities were hardly involved except as spectators. During the demonstration period, the more active involvement of the vendor and the users emerged, although government participation was still important in supporting continuing R&D and underwriting the demonstration reactors. By the time the dissemination period was reached, the government R&D activities had largely ceased as the vendor and consumer assumed the primary development and dissemination responsibilities. The national nuclear reactor developmental programs may be seen as involving a number of different organizations with their particular objectives; furthermore, these institutions and objectives and their ordering changed during the three RD&D and dissemination phases.

Figure 8-1 graphically assesses the relative importance of the nuclear reactor technology development objectives drawn from the six national reactor development programs. The charts depict the different objectives as they changed during the course of the three developmental periods. Figure 8-1 is especially instructive for appreciating the changing hierarchy of the major institutional actors during the development as reflected by the changing relationship of the multiple objectives. In the development period, the science for science's sake and national prestige objectives, both largely sponsored by the government developer, were the

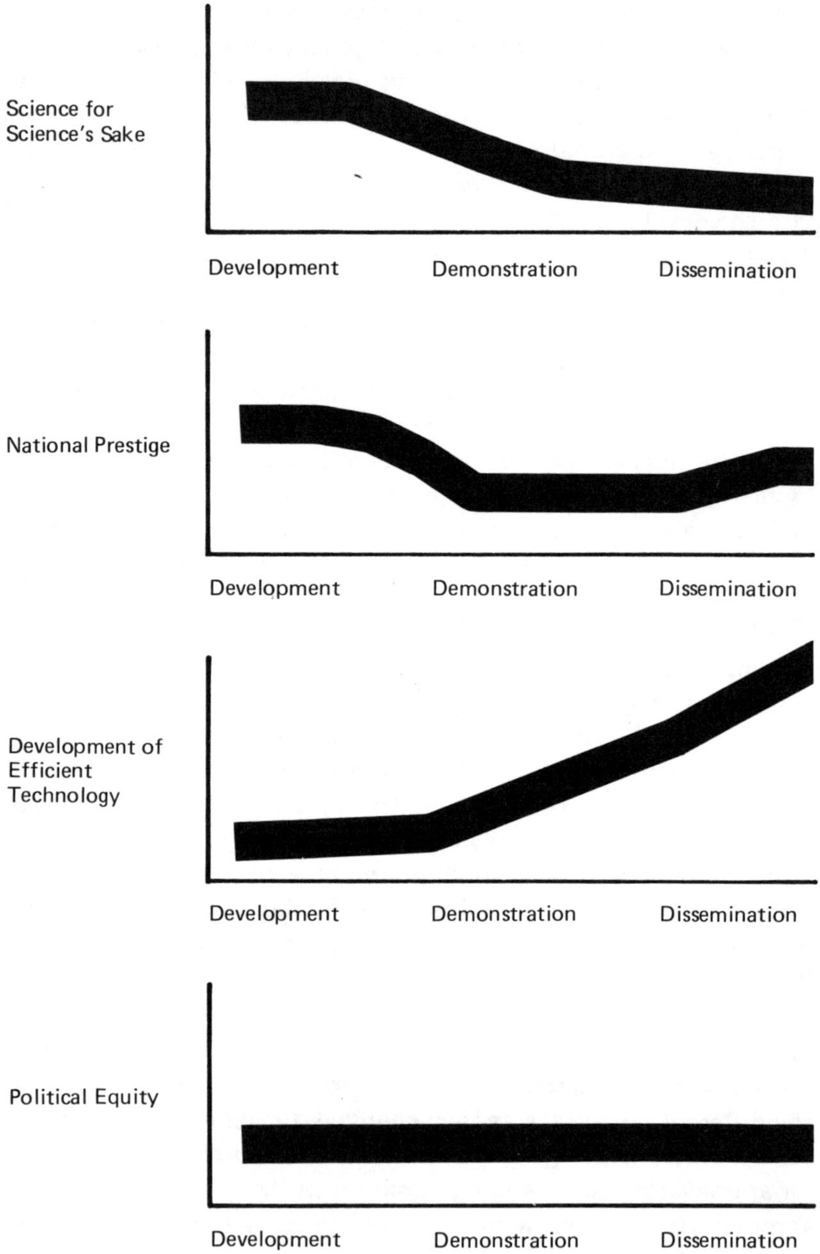

Figure 8-1. Relative Trends of Objectives for Technology Developments

primary objectives influencing the developments. The major exception to this general observation is England, in which the export market for reactors was an early consideration by the government for supporting a faltering balance of trade. A minor exception is West Germany, where the initial impetus for developing nuclear reactors was provided by a combination of German industry and electric utilities, but government spokesmen issued strong statements regarding Germany's need to develop nuclear energy as a symbol of German industry and to reestablish Germany's scientific research base.[7] For the most part, however, Figure 8-1 displays the predominant role during the development period played by the particular objectives of the government R&D agencies.

During the demonstration period, the development objectives seen in the case studies changed in their relative ranking and generally manifested a transition from the developer's objectives to those more characteristic of the vendor and consumer institutions. In every nation studied, the efficient technology objective, which is linked to the vendor and consumer actors, became more important in the national ordering. It is still possible to see the influence of the government development agency as national prestige remained an important objective throughout this period. For instance, Allen argues that in the United States the Eisenhower Atoms for Peace plan (1954) was little more than propaganda to garner international goodwill and prestige.[8] Hogerton writes that "the thrust of [American] accelerated nuclear power program was not sparked by a specific technical development, nor did it develop solely from the building of economic pressure within the U.S. power industry. Considerations of *national prestige* and foreign policy were of central importance."[9] The British and French nuclear development agencies were still engaged in their exclusively domestic reactor research and effectively excluded their vendor and consumer

---

7 OECD, *Changing Priorities for Government R&D* (Paris: OECD, 1973), documents the strong support given by the West German government to "General Advancement of Science;" cited by Keck, "West German Science Policy Since the Early 1960's," pp. 147, 151.

8 Allen, *Nuclear Reactors for Generating Electricity*, pp. 31-32, 76-77.

9 Hogerton, "The Arrival of Nuclear Power," p. 23; emphasis added.

organizations from the technology development processes. Toward the end of the demonstration period, both nations' governments explicitly rejected the bids by their nationalized electric utilities and electrical construction industries that the American LWR be imported. Indeed, both governments cited remarkably similar expressions of patriotic faith in their indigenous reactor systems.

Finally, during the dissemination period, the primary TDS actor became the consumer, with the vendor slipping to a supporting role and the government R&D organizations largely having moved to other projects; whatever government influence remained was in regulation (e.g., designing and enforcing safety and environmental standards). Because the users in these cases were primarily concerned with the low cost delivery of electricity, efficient technology was the most important objective, with the vendor's concern for expanding his sales reflected in a growing interest in export. National prestige still lingered as an influential objective; but in most cases, it was clearly subordinate.

Although this pattern generally was seen, the two discrepancies merit discussion. The British AEA found itself in a situation where its second-generation reactor, the indigenously developed AGR, was proving too costly and technologically inefficient to warrant continued dissemination. The alternative reactor choices once again were narrowed to the U.S. LWR and the British-developed (with Canadian support) SGHWR, with the U.K. AEA being principally responsible for the evaluation. The British cabinet announced its selection of the SGHWR because, according to the accompanying White Paper, the "SGHWR offers particular scope for British nuclear technology."[10] This decision stood in direct opposition to the stated preferences of the CEGB (the consumer) and the single remaining industrial nuclear consortium (vendor), both of whom strongly favored the Westinghouse PWR. As one reporter observed, the Labour government and AEA simply refused to "sell short on the nuclear technology in which all British governments have invested heavily since World

---

[10] Quoted in Walsh, "British Choose Own Reactor for Nuclear Power Program," p. 511.

War II."[11] Consideration for British prestige—some called it "nuclear nationalism"[12]—still continued to play a predominant role in the British development calculus. Writing about the British program of early 1970s, Wonder comments that "Extrinsic and random factors such as prestige and personality were just as important as the intrinsic techno-economic merits of different reactors or industrial structures."[13]

The analogous French situation saw an entirely different outcome, one more consonant with the generally observed pattern. The French nationalized utility argued against CEA for the more efficient U.S. nuclear technology; the French nuclear construction industry contended that the U.S. reactor technology, even under license, could provide France with a more attractive export option than the French gas-graphite reactor system. Although the minutes of the debate are not available, one can speculate that the French government decided that continued support of an inefficient technology was in fact damaging French prestige while undercutting EdF and the projected French nuclear industry. Rather than reaffirm its backing of CEA and its reactor preference, the government chose to support the vendor and consumer preferences. Perhaps it considered that whatever prestige might have been lost by adopting the American reactor system could be regained with an aggressive merchandising program, a strategy that would have been underpinned by the French perception of the West German example.

If the description of the changing priorities of national objectives and their constituent organizations is correct, the important conclusion can be drawn that the development and dissemination of nuclear power reactors was less a case of a developing technology and more of interorganizational conflict. The development and dissemination of power reactors was uneventful in nations where the appropriate institutional actors exercised leadership during a given phase of the development and the transitions between the three phases were fairly smooth (i.e.,

---

[11] Ibid.

[12] Ibid., and Wonder, "Decision-making and Reorganization of the British Nuclear Industry."

[13] Ibid., p. 266.

one found an "integrated" TDS). This was true in the United States, Canada, probably in the Soviet Union, and, to a lesser degree, in West Germany. Serious problems occurred in the development programs of Great Britain and France where an institutional actor retained control of the program after a developmental transition (i.e., after an institutional transition should have occurred).

This is not meant to suggest that the development and dissemination of power reactors were entirely political exercises. Clearly, there were important technical obstacles that had to be overcome, but in most cases, these appeared to be more manageable than the organizational problems that beset and inhibited even the American and Canadian programs.[14] The German experience was hindered by the reticence of the central government to become involved; it was not until the vendor and consumer actors forced more active *Bund* participation that the West German TDS achieved the degree of institutional integration that encouraged the expansion of the German nuclear programs. The major lesson is that there are discernible developmental periods during the process of technology development and dissemination. Each is generally defined in particular requirements and expectations, and each is characterized by a set of attributes and an ordering of objectives.

Policy Observations

The most important policy observation that can be made from the six national case studies in the perspective of this analytic framework derives from two fundamental findings. First, different organizations and their constituent objectives are particularly critical or predominant at certain stages of the development and dissemination processes. Second, different attributes are especially important for the pursuit of different objectives; for instance, an integrated TDS is central to the efficient

[14] Allen, *Nuclear Reactors for Generating Electricity*, contends that misunderstandings between the AFC and the American vendors and consumers were a major obstacle in the U.S. nuclear reactor development program.

technology objective but not if science for science's sake or national prestige were the motivating objectives. It is essential to recognize the stage of the development process and to emphasize the appropriate objectives and set of attributes. Concomitantly, objectives with certain short-term payoffs can have high costs in the long term, even if the persons or organizations that gained the benefits are no longer in position to pay the costs. For example, a nation might be ill-advised to stress the TDS attribute and its efficient technology objective during the early stages of a development. Scientific research is inexpensive compared with the later costs of demonstration and dissemination. To worry about the commercialization problems or technical efficiency when the technology is in its infancy is premature at best and disconcerting at worst. By the same token, the science for science's sake objective and the government's development activities should not be principal concerns at the time of the decision to accept or reject a demonstrated technology. By that stage, the existence of an active scientific community is much less important than earlier in the development of that technology, and the government's role should have been reduced to providing operating subsidies or altered to that of regulator.

This requirement for period recognition is not as intuitively obvious as it might appear. The early British and French programs were motivated to attain a working reactor system as quickly as possible because the British needed to develop the export market and both nations wished to overcome a perceived energy shortage. The emphasis on these goals during the development period caused them to stress objectives and attributes of the development that were inappropriate for that stage. A similar observation may be made regarding the apparent English decision to place national prestige above efficient technology during the later stages of the British reactor development program.

Technology development represents a number of different objectives as it progresses over time. A single evaluative metric for a developmental program is inadequate and potentially misleading.[15] For instance, if the indigenous development of an efficient nuclear reactor technology were the criterion upon which

---

[15] The problem of assessing programs with changing measures and at different

the French program were judged, the French gas-graphite reactor development must be judged a failure. But if the criterion were the selection and production of an efficient reactor, the current and projected number of domestic and exported French reactors belies the first assessment.[16] By the latter criterion, the French program is a success, although a late bloomer. Similarly, arguments can be posed that the French reactor program has added to the national prestige, benefited the French scientific community, and served as a valuable export commodity during the 1970s, all of which undermine the evaluation of the French reactor development program by a single objective. The policy implication is clear. A technology development must be measured against a number of objectives, and the ultimate assessment of its "success" or "failure" must reflect that multiplicity of objectives. Furthermore, just as economists have alerted the policy community to the possibility of unintended externalities that occur once a policy has been carried out,[17] the existence of multiple objectives should alert policy researchers to the idea that technology developments might be responding to objectives not included in the analyst's list.

Objectives and Attributes

A review of the cases suggests which attributes were most influential in determining the "success" or "failure" of the individual objectives, but the limited number of cases, the idiosyn-

---

points is discussed by Greenberg et al., "Developing Public Policy Theory," pp. 1536-1538.

16 Walske, "Nuclear Electric Power . . . ," p. 47, credits France with more nuclear reactors operating, under construction, ordered, and planned than any other nation except the United States. Robert J. Samuelson, "Pulling the Plug on Plutonium," *National Journal* 9, 31 (July 30, 1977):1178, cites Department of Energy figures projecting France relying on nuclear reactors for 90 percent of its electricity supply by the year 2000; although a footnote alerts the reader that the projections may be "unreliable," the estimated French dependence is much greater than any other nation.

17 See Eads, "U.S. Government Support for Civilian Technology," for a discussion on the policy "externalities" regarding the fate of the economic theory of externalities when placed in the hands of politicians.

crasies of the technology and the institutions, and the qualitative nature of the data render formal policy recommendations premature. Rather, a number of *observations* relevant to policy may be extracted from the objectives and attributes matrix of the analytic framework.

Scientific Research for Science's Sake. The key attribute in defining and supporting this objective was the *active scientific community*, which stood the most to gain from a national program. The case studies highlight two less obvious relationships. First, this objective requires a low level of resources invested, even if the cost of constructing new research facilities is included. Therefore, the national government can honor this objective at a low level of expenditure. Second, even if science for science's sake usually suggests scientific research on the frontiers of an area, which implies the importance of being an *early starter* so as to assure one's position in the technological vanguard, such a relationship is not a requirement. The West German government heavily emphasized the general advancement of nuclear science even though German atomic scientists were not able to initiate their research until the mid-1950s, a full ten years after the other nations studied. More to the point, German scientists made important scientific contributions.

Finally, whatever contributions a *concomitant military program* might make to a civilian technology development, they are the most important during the early development period and to science for science's sake. Military technology is usually more advanced than its civilian counterpart, so whatever spillovers occur would be during the earliest, most formative stages of the civilian development while it is still exploring the state of the technology. During the later stages, civilian technology is more concerned with reliability and costs than with advancing the technology. As Eads and Nelson pointed out, the conventional belief that the Boeing 707 technology grew directly from Boeing's military R&D contracts is not supported by the evidence.[18] However, the nexus between military and civilian nuclear research might be particularly and uniquely intimate; more cases need to be ex-

---

[18] Eads and Nelson, "Government Support of Advanced Civilian Technology."

amined before the general role of military research can be understood in terms of benefits to civilian technology.

National Prestige. At best, national prestige is an elusive objective from both the definitional and attainment standpoints. It is difficult to state which attributes were the most significant in contributing to its successful achievement. However, the *resources invested* in a given technology have an interesting effect on a development if national prestige is an objective. During the early stages of a development, a small government investment to support scientific research and build test reactors could result in a large boost to national prestige. However, this investment can have the effect of tying the national prestige to the technology development, thereby making it more difficult for the government to abandon the project should it later prove technically or financially infeasible. The early national investments for establishing (or maintaining) a nation's prestige were quite low, but as the development continued, the resources necessary for the maintenance of the national prestige objective became formidable. The political sunk costs incurred by the national prestige objective outweighed the economic calculus of sunk costs. The reluctance of the British government to terminate the British reactor development program, partly because of a feared loss of national prestige, certainly has proved extremely expensive. The policy implication is that a seemingly low investment in a technology development for purposes of national prestige can grow into a very costly investment if that objective maintains a predominant position.

Development of Efficient Technology. The appropriate time for this objective to become critical is during the later stages of the demonstration period and the dissemination period. Therefore, the role of the *active scientific community* is minor for this objective. The issues are no longer those of the advancement of science; reliability, cost, and technical considerations now predominate.

This study does not support the value often ascribed to a *concomitant military program* in achieving an efficient technology. Two nations that had ongoing atomic military programs (the

United States and the Soviet Union) successfully developed an efficient civilian reactor technology, but similar successes were gained by Canada and West Germany, which had no military nuclear research program. Finally, the concomitant French and British military research programs did not save those nations from experiencing great difficulty in attaining an efficient reactor system.

Conventional wisdom has long held that the concurrent development of *multiple technological options* brings significant benefits to a development program, especially for the development of an efficient technology, because the presence of multiple options permits the selection of the best alternative. The six national nuclear experiences certainly do not refute this. The multiple reactor options attribute was a significant characteristic of the programs that emphasized the efficient technology objective. The nations that were generally considered more "successful" (Germany, the United States, and the Soviet Union) developed multiple reactor options, and the less successful nations developed only one (France and England).[19] Upon reflection, however, the nuclear reactor developmental experiences certainly do not confirm the thesis. Canada developed a single, efficient reactor technology; the CANDU reactor, developed on a limited scale, competed in the international market with greater success than the British and French models.

The development of multiple reactors in the United States and the Soviet Union does not confirm the hypothesis that these nations developed an efficient technology because their decision-makers could choose among a number of different reactor designs. Perry raises the possibility that had the AEC decided early on to second Rickover's choice of reactors and focus its slow fission reactor development energies upon the PWR (and possibly the BWR), it would have not been diverted by the multiple competing technologies, and commercialization would have been achieved much earlier than the late 1960s.[20] If multiple

---

19 See Joskow, "Research and Development Strategies for Nuclear Power in the United Kingdom, France, and Germany"; and Burn, *The Political Economy of Nuclear Energy*, for such evaluations.

20 Perry et al., *The Light Water Reactor*, pp. 82-83.

technological options were not the key to an efficient technology, the British and French difficulties were not so much caused by a faulty R&D strategy as by making the wrong choice. Again, the situations were much too complex to permit so easy a judgment; given their resource and political constraints, neither nation made the "wrong" choice, for the decision was more of a Hobson's choice. Regardless, the point to be made here is that one cannot confidently accept or reject the multiple technology options hypothesis on the basis of the nuclear reactor development programs.

The presence of an *integrated TDS* was central in the development and dissemination of an efficient technology. The active participation of all the TDS actors too early in the developmental process was unnecessary or even potentially harmful in terms of undermining the technology's development—for example, the consumer's emphases on costs and reliability might have been counterproductive during the early development stages; or a core meltdown in a utility company reactor that had been built during the early demonstration stage would have cast a dampening effect over the subsequent development. This finding supports the institutional concerns voiced by Lambright and Teich as necessary for the dissemination of a civilian technology[21] and with the analysis presented by Baer and his colleagues.[22] Unlike them, it differentiates among the actors and their most effective roles at different times during the RD&D and dissemination processes.

This does not mean that the developer, vendor, and consumer are the only three important institutional actors. Labor unions and environmental lobbies, among others, can have a pivotal role in a development. What it does argue is that the developing agency, especially if it is a government agency free from market considerations, should pay close and early attention to the vendor and consumer and how their efforts and requirements might be integrated into the developmental process at the appropriate stages. This directly confronts the issue of how the developed technology will be disseminated for civilian consumption.

---

[21] Lambright and Teich, "Technology Transfers as a Problem. . . ."

[22] Baer et al., "Government Sponsored Demonstrations of New Technology."

Typically, the government has not been well suited for the dissemination role. Therefore, early attention to the requirements of the vendor and consumer actors is very important.

Two observations may be made about the *resources invested* attribute as it affected the efficient technology objectives. First, the resource demands are not of the same magnitude for each of the three developmental periods nor the same for the different actors during the period. The early development stages were much less expensive than the later demonstration and dissemination periods. This observation implies that "buying in" should not be considered an inexpensive means of obtaining a technology. Even after acquiring most of the technical developments on license, the West German nuclear research program languished in the mid-1960s until the federal government increased its contributions more than 250 percent.[23] The American commercialization experience points out how expensive the dissemination of efficient technology can be; Westinghouse and General Electric, the two major U.S. reactor construction firms, are estimated to have lost more than $750 million on their twelve turnkey reactors by grieviously underestimating the cost and technology uncertainties.[24]

Second, the successful development and dissemination of an efficient technology were largely determined by the cooperative involvement of the developer, the vendor, and the consumer. For nuclear reactors, it was particularly important that the integrated TDS be present. To help ensure the continued participation of the vendor and consumer, the U.S. government insisted that they invest their internal resources in the development (i.e., incur sunk costs). There are many indications during this protracted period that a given investor's attitude toward risk may vary, even when there are no objectively perceived changes in the prospects for success. In such cases, the amount of resources invested may act as a lid or damper on risk adverse behavior, without which the

---

[23] OECD, *Changing Priorities for Government R&D*, lists a change rate in German government spending on nuclear research of 2.58 between 1961 and 1967; by comparison, the equivalent figures for France, the United States, and England are 1.88, 1.15, and 0.78. Cited in Keck, "West German Science Policy Since the Early 1960's," p. 195.

[24] Perry et al., *The Light Water Reactor*, p. 99.

development might be terminated. As posited in Chapter 2, these sunk costs created a commitment to or momentum behind the continued vendor and consumer support of the program that might otherwise have been lagging, especially as the reactor programs reached the expensive development and dissemination stages.

Nations whose vendor and consumer industries had little or no requirement to contribute their R&D resources, specifically France and Britain, were unable to develop an efficient technology. Some observers have argued that this failure was at least partly due to the restricted participation of the French and British utilities and vendors.[25] Wonder argues that the British consortia had little cause to improve their technical capabilities because each member was assured an equal share of the reactor construction program by the rotating, noncompetitive system of allocating reactor orders, or what was referred to as "Buggin's turn."[26] Nor did they have the opportunity to commit their research skills because of the effective AEA monopoly on atomic research. These policies sharply reduced the research investment and participation incentives of the consortia members,[27] thereby reducing their potential for identifying and avoiding nuclear technology that would be inefficient and effectively undermining the integrated TDS needed to deliver the efficient nuclear technology.

The U.S., Canadian, and West German reactor development programs stand in contrast. The AEC insistence that American utilities invest their money in reactor technology under the PRDPs provided the utilities with a strong vested interest for the development of an efficient technology beyond the preference predicted by the theory of the firm for the most cost-effective producer of electricity. Similarly, early investments by West German vendors and utilities promised their continued interest. Contrarily, national vendor and consumer industries with few

---

25 Joskow, "Research and Development Strategies for Nuclear Power in the United Kingdom, France, and Germany," and Burn, *The Political Economy of Nuclear Energy*.

26 Wonder, "Decision-Making and Reorganization of the British Nuclear Power Industry," p. 265.

27 Burn, *The Political Economy of Nuclear Energy*.

resources invested had less of an incentive to attain an efficient technology. In the face of technological and investment uncertainties, the organizational commitment represented by such costs took precedence over the calculus of marginal costs and provided at least partial reason to continue the development despite cost overruns and schedule slippages. Assuming the experience of the development and dissemination of nuclear power reactors is representative of the broader class of civilian technology developments, the public policy implication is that if a nation's decisionmakers decide that they wish to develop an efficient civilian technology, vendor and consumer financial participation is important for ensuring their continued involvement and the productiveness of the TDS.

Given this behavior, one must ask what level of prior resources invested are necessary to maintain the developmental commitment of the organization. This assessment obviously presents some semantic problems, specifically, "how much is enough?" No definitive answer can be given for the cases studied; no national program appeared to founder principally for lack of funding. It is impossible to distinguish the development of efficient technologies by the level of resources invested. Burn claims that the British AEA invested an equivalent amount of resources in the British reactor development program as the U.S. AEC did for the American program, but the end results clearly differed;[28] the Canadian program surely cost less than the French development. Unfortunately, the noted difficulties in ascertaining program costs and the variability of national circumstances and organizational behavior (especially over time) make it impossible to specify from this limited number of cases what the appropriate investment threshold might be to ensure a continuation of the development commitment.

Chapter 2 proposed that one reason a government might sponsor RD&D toward an efficient technology would be to develop an attractive export commodity that could benefit a given industrial sector and the national balance of payments. This goal was explicitly stated by all the nations studied here, and by England and France from the early stages of the development period. In

---

[28] Ibid.

the competitive marketplace, the ability of a nation to export its reactor system should be largely a function of the cost and technical efficiency of that system relative to alternatives (e.g., the LWR versus the gas-graphite reactor system). This could conveniently be measured by the number of reactors sold outside a nation's borders. However, the presence of nonefficiency criteria, such as national subsidies of reactor exports, licensing arrangements, technical perquisites (e.g., reprocessing plants), political domination (e.g., the Soviet bloc nations), and in some cases, hidden payments, all distort the validity of this simple market test. It is not possible to ascertain how the desire to develop an efficient technology for export purposes affected the developmental process because too many nonefficiency factors enter into the export calculus. However, a number of observations may be made about technology exports in terms of the attributes.

Perhaps the most interesting insight is that the putative value of an *early starter* in the development of an exportable reactor was not supported by the nuclear reactor development data. The British program, which was an early starter and had explicit export goals, fell short of attaining this goal; the West German industry, a later starter, achieved notable success. Similarly, the French nuclear export industry can be viewed as a late starter because it did not adopt the exportable U.S. LWR until the late 1960s; since then, it has been fairly successful in attaining its share of the reactor export market[29] (see Table 8-1). The policy implication is that, other considerations being roughly equal, later production skills are more important for developing an exportable technology than being in the technological vanguard. Indeed, too early an emphasis on the export market might have the unfortunate consequence of binding a nation's industry to an outdated technology, a condition that occurred in England and threatened in France. Obviously, being on the technological frontier is not always a disadvantage, but the nuclear evidence suggests that it is not unambiguously favorable. A second policy im-

---

[29] Ann MacLachlan, "Vive La France Nucleaire!" *Energy Daily* 6, 199 (October 16, 1978):3, recently wrote that "They don't know quite how it happened that way, but the French today find themselves increasingly at the forefront of world nuclear power development."

Table 8-1. Export Orders for Nuclear Reactors

	U.S.	*FRG*	*France*	*Canada*	*Sweden*	*USSR*	*U.K.*
Total exported reactors	66	10	7	5	2	22[a]	2
Total domestic reactors[b]	224	27	40	20	10	25	38
Percent export/all reactor sales	22.7	27.0	14.9	20.0	16.7	46.8	5.0

[a]All Soviet reactor exports are to Communist bloc nations.

[b]The reactor sales cited by Samuelson do not match those of Walske, "Nuclear Electric Power . . . ," p. 97.

Source: Samuelson, "Pulling the Plug on Plutonium," p. 1185, citing Westinghouse Electric as his source.

plication is that early advantages cannot always be maintained and attention should be paid to export requirements in all periods of the development if the technology is meant to be an export commodity.

Again, the value of the *active scientific community* is less important here than seen previously because the emphasis is less on the advancement of science and more on the efficiency of the technology. The developmental emphasis has moved past the esoteric toward the reliable, particularly if the reactor is being purchased by a nation whose technical resources cannot match those of the exporter.

A final attribute that appeared to be influential in determining a nation's ability to develop an export market for reactor systems is the *integrated TDS*. The evidence suggests that the selling of reactors beyond one's national boundaries required the close cooperation of all three actors. The domestic consumers must appear to the world market as sufficiently satisfied with the product; few nations want another's technical refuse. The vendor and the national government must work closely to secure foreign reactor sales; both are necessary and neither is singularly sufficient. The British AEA was unable to export atomic reactors partly because the English reactor program lacked the necessary industrial base and its merchandising expertise. The West German and French governments supported their nuclear industries' activities abroad, even permitting previously restricted technology to be included so that the sale could be consummated. The American reactor export efforts were largely conducted by electrical construction firms and have been the most successful in terms of reactors sold; however, they would not have succeeded had not the U.S. government made enriched uranium available through the Atoms for Peace program and contributed close to $4 billion in direct loans and guarantees.[30] The export TDS is somewhat different from that earlier formulated for the efficiency objective because the government actor is now much less a technology developing organization and much more a technology subsidization organization with different motivations and possibly represented by different institutions.[31]

Political Equity. As originally defined, the political equity objective was meant to display the influence of political interest groups on the development and dissemination of nuclear power reactors. During the early stages of the development process, the *active scientific community*, acting largely out of group self-interest, was able to persuade political decisionmakers that their group fully warranted government funding; political equity was extended to the support of the scientific research community. The attributes that most clearly reflected the interactions and changing contexts of the pertinent interest groups (although on an aggregated basis) were *multiple technological options*, and the *integrated TDS*. The first represents the presence of a number of competing technical groups, each advancing its particular design, and the funding agency allocating its resources among those groups. This suggests that multiple technical alternatives were not solely being proposed as a hedge against technical difficulties; rather, they can be depicted as competing interests. Perry speculates that the multiple reactors de-

---

30 Direct U.S. Government loans to nonindustrialized nations equal $2.4 billion and guarantees equal another $1.4 billion, for a total of $3.8 billion. Data from Export-Import Bank and DOE; cited in Samuelson, "Pulling the Plug on Plutonium," p. 1181.

31 A recent example of how government and industry export objectives could be in opposition is reported by Youssef M. Ibrahim, "U.S. Nuclear Industry Sees Bill as Threat to Exports," *New York Times* (February 15, 1978), pp. 51, 57.

signed and developed within the United States were more a function of institutional demands to participate in the new technology than technological uncertainties.[32] Similar logic seems applicable to the Soviet reactor development program. The organizational conflicts among the TDS actors manifested the struggle by the three major institutional actors to gain control of, or at least participation in, the reactor RD&D deliberations. The lesser the degree of integration within a nation, the more heated those conflicts became until, in the cases of the British and French programs, the struggles had to be resolved in the highest councils of government.

Supply and Demand

A final policy implication—the need to match the development of technologies of this magnitude to the appropriate market demand—can be extracted from this analysis. Some markets are more responsive to governmental fiat than others; however, these markets are rarely sufficient for the financial self-sufficiency of the relevant sectors of the private economy or for the government to recoup its investments in developments of this size. It was clear that the demand of the British domestic market for nuclear reactors could not have supported four or five nuclear construction consortia. Therefore, AEA had to predicate the viability of the English reactor industry on the export market if it were to be sustained. Unfortunately for the English reactor vendors, the international market was much more insulated from the pressures of Whitehall than CEGB. Had there been an accurate estimation of the demand requirements or those the government could at least affect, the disparities between what the domestic market could absorb and the number of units necessary to sustain the industry would have revealed an excessive reliance on the much more problematic international marketplace. Of course, these estimates might be neglected if considerations of national prestige remained a driving objective into the demonstration and dissemination stages or the government chose to ignore the fact of limited markets for its own political or social reasons.

32 Perry et al., *The Light Water Reactor*, pp. 82-83.

However, the evidence of the nuclear reactor programs strongly argues that few accurate estimates were made of the predicted demand structure for electricity, a shortcoming that originally impelled the British and French to initiate their nuclear reactor programs and later to expand them in the face of a declining market.

Evaluating the Analytic Framework

We can now address the third question raised at the beginning of this chapter, what is the value of the proposed analytic framework from the intellectual and policy perspectives? In both instances, the evaluation is guardedly positive.

The most important insight offered by the analytic framework for both the intellectual and policy perspectives of the reactor development programs is the explicit recognition of the multiple common objectives that shaped and guided the different programs. This, of course, implies the existence of multiple organizations that all have an effect on the technology development programs.[33] One should also recognize the competitive and dynamic natures of these multiple objectives and how their interactions defined the reactor development programs. In a political environment in which resources were limited, the different organizations and objectives were forced to compete against each other for priority and the attendant resources. Furthermore, this was not a one-time encounter. As the political environs changed and the various attributes acquired or lost importance, different organizations and their particular objectives assumed new positions in the development hierarchy. Those nations whose objectives and organizations appeared to make a fairly smooth transition in the face of these dynamics had fewer developmental problems than those in which the friction of organizational transitions seemingly denied the changing political, economic, and technological conditions.

From the intellectual or heuristic standpoint, the framework

---

[33] The latter observation is the crux of Allison's Models II and III, *The Essence of Decision*.

provides a concept and tool for coherently explaining a technological development in terms of the political, institutional, and economic forces that shape (either positively or negatively) a given technology. By itself, this is not unusual; a large number of national case studies have illuminated similar themes. However, in the context of the multinational developments of the magnitude and complexity encountered here, the framework's contribution is unique and valuable. The objectives/attributes matrix synthesizes a large amount of data into a single analytic structure. Even with the necessary reservations, evidence from the development of different technical programs and different nations can be juxtaposed and compared to obtain a general understanding of what technical and social pressures affected the development and dissemination of a civilian technology. The capability to fit disparate information into a common framework that renders plausible explanations for the entire range of the studied phenomena, therefore, is probably the most important intellectual accomplishment of the analytic framework for technology development.

A second feature of the framework as a heuristic tool is its ability to identify what information or data the research is lacking, much as Kaplan argues it should.[34] The comparative feature of the construct encourages greater confidence in this regard than the case study mode. Noting how the data missing from one study enrich the other studies, the comparative analyst can discern overall patterns of data deficiencies. For example, the political equity objective, which is amply supported in the political science literature, was observed to be of only marginal importance in the ordering of objectives, which did not conform to expectations. One could logically argue that either the expectations were amiss, or more likely, the appropriate data were not available. If the latter case is valid, the shortcoming can probably be addressed in later iterations.

A final, more tentative advantage of the model is that it appears to have (or could develop) some predictive possibilities. If the model is used to interpret the data, certain predictions might be cautiously advanced. For instance, a nation's development of

[34] Kaplan, *Conduct of Inquiry*, p. 271.

thermonuclear or breeder reactor technologies might be predicted to be problematic if its early demonstration reactors were held strictly accountable to a cost efficiency criterion. The Soviet reactor development authorities specifically rejected the relevance of this criterion during the demonstration period.[35] Similarly, a technology whose dissemination decision was imminent but still dependent on a concurrent military program for its technology could be seen as suspect, in the same way as a technology whose ability to attain a commercially cost effective operation heavily relied upon attaining an export market. These modest predictive claims, of course, require tempering and validation through further testing. They also are a key to the second area for evaluation, the policy worth of the analytic framework.

The framework's policy value must be judged by its capability to identify and extract policy observations out of an unstructured body of information and against its capacity to minimize mistaken interpretations. Again, it appears to have been quite useful for at least three reasons. First, as defined, it permits comparisons of multiple experiences, thereby reducing the limitations inherent in a single case study. This capability protects the policy scholar from biasing his or her recommendations on a case study of a technology development that might be seen as anomalous in a later light. Just as important, it permits the policy analyst to confirm a finding that might otherwise appear as unduly dependent on a single study. For example, the contention that the British Comet jet liner suffered from premature commercialization might be dismissed by skeptics who claim that other factors were more central to its failure and that one should not generalize from one example. However, the later failure of the British Magnox reactor compared with the U.S. LWR gives added substance to the observation that the early commercialization of advanced technology might be a mixed blessing. In short, the analyst can benefit from the advantages

---

[35] Petrosyants, *From Scientific Search to Atomic Industry*, p. 115, explains "that the first power unit of the Novovoronezh NNP, and the whole power plant, was constructed not only as a commercial complex, but as a demonstration facility as well."

that accrue from using a larger data set, advantages that formerly were difficult to obtain and tenuous when multiple cases were attempted.

A second policy benefit of the general model is that it identifies separate developmental stages and highlights their requirements. This forces the policy adviser to recognize that there are different periods in a technology's RD&D and dissemination, each with its own requirements, and permits him or her to recommend measures appropriate to those requirements. The framework offers some policy insights into what are the appropriate objectives and attributes of a development at a given stage of the RD&D and dissemination processes.

As noted above, a third policy feature of the analytic structure is that it stressed an analytic level particularly useful to the policymaker working on large technology developments. To perceive or treat the nation as a unitary decisionmaking entity (a "black box") does not give the policymaker much insight as to what levers he or she might adjust to direct a development program. The implementation problems are obscured because the policymaker cannot understand which organizational units are responsible for his or her policy or the interinstitutional conflicts whose tugging and pulling defines the developments program. These blinders severely limit the ability to design and carry out a chosen policy. An appreciation of the institutional forces permits the policymaker to initiate and evaluate the policies and programs with greater confidence than if the other two analytic levels were subscribed to. In brief, the analytic framework gives the decisionmaker not only information about what the proper objectives might be at any given time but also how best to meet those objectives.

A final policy benefit of the framework is that it emphasizes the institutional elements of a technology development rather than the scientific or technical aspects. The evidence to date suggests that it is extremely difficult to expedite scientific research or technological developments; such efforts have often resulted in significant cost overruns and technical deficiencies. It is more possible to encourage a development by removing or at least reducing the institutional obstacles that might otherwise impede the development and dissemination of a technology—policies that can be instituted by the policymaker—than to expedite technical development itself, which is more problematic.

These evaluations of the intellectual and policy aspects of the analytic framework do not, of course, mean that it is in its ultimate form. As discussed in Chapter 2, it was proposed as an initial attempt to understand a series of complex developments, to structure a large amount of varied information, and to describe, in skeletal ways, the logic of RD&D programs and institutions. In its current form the construct is more suggestive than conclusive. An additional task would be to offer new directions for future research. The limitations contained in the definitions, the data acquisition problems, and the restricted number of cases all indicate that it is still in its postulated form and that important work is yet to be done before it can be considered a policy analytic tool applicable to a broader range of technology development.

9

A Brief Comparison of the Development and Dissemination of Government-Sponsored Civilian Technologies

This chapter compares nuclear reactor development experiences with similar programs of civilian technology developments and explicitly indicates what further research would be necessary to test a general model of government-sponsored civilian technology development and diffusion.

A Comparison of Technology Developments

A general "model," as currently used in the social sciences, is a series of logical, interconnected sets of observations codified into rules or laws that can satisfactorily explain all the phenomena within the stated purview. For an issue as protean and complex as technology development and the component technological, political, economic, and societal forces that shape and guide a development, any postulation of a definitive model at this time would almost certainly be premature.[1] Too early and dogmatic a resort to formal model specification could be counterproductive. Kaplan cautions, "We tinker with the model when we might be better occupied with the subject-matter itself."[2] Furthermore, a model that is not explicitly oriented toward policy in its thrust and construction is not likely to have any direct policy applications or insights.[3] Indeed, the applications of nonpolicy models to public policy issues can be dangerously wide of the mark, thereby rendering the policy ineffective or, in some cases, exacerbating the problem.[4]

For these reasons, an exercise in formal model building has been deliberately avoided. What has been presented is an analytic framework and taxonomy that permitted a survey of six different nations' programs to develop and diffuse nuclear power reactors. The framework was not meant to serve as a theory, but it does highlight the context of institutional behavior and the main patterns of their interactions. Rendering an area intelligible is not the same as offering an explanatory or predictive theory. Still, as suggested in Chapter 8, the framework does reveal some observations relevant to policy and, if carefully manipulated, perhaps offers some longer term rewards to the policy scientist. As Lowi contends:

> Finding different manifestations or types of a given phenomenon is the beginning of orderly control and prediction. . . . Moreover, to find

---

[1] Rosenau proposed a "pre-theory" stage; see James N. Rosenau, "Pre-theories and Theories of Foreign Policy," in R. Barry Farrell, ed., *Approaches to Comparative and International Politics* (Evanston, Ill.: Northwestern University Press, 1966), esp. p. 40.

[2] Kaplan, *The Conduct of Inquiry*, p. 279. Also see Albert O. Hirschman, "The Search for Paradigms as a Hindrance to Understanding," *World Politics* 22, 3, (April 1970):329, who writes, "The theoretical quick fix has taken its place in our culture alongside the quick technical fix."

[3] Peter deLeon, "Public Policy and Political Development: The Case of the SSRC," *Policy Studies Journal* 5, Special Issue (June 1977):627-628.

[4] Brewer describes how the Arthur D. Little model of the San Francisco housing market was built by a physicist and, not surprisingly, compared rental prices with the physical phenomenon of hysteresis. Brewer comments: "If a model builder has never been sensitized to the details of a specific empirical context, one should not find faults with his inferential leaps . . . from expanding and collapsing magnetic fields to expanding and collapsing rentals." Garry D. Brewer, *Politicians, Bureaucrats, and the Consultant* (New York: Basic Books, 1973), p. 143.

the *basis* for classification reveals the hidden meanings and significance of the phenomenon, suggesting what the important hypotheses ought to be concerned with. This is precisely what a policy taxonomy might do for the study of politics.[5]

Rather than advance a modest model at the time, it would be more constructive to ask if the patterns of institutional interactions observed in the six national reactor development programs are consonant with other large-scale government-sponsored civilian technology development programs. Are the patterns of behavior seen in the national reactor development experiences found in other technology development areas (do the propositions appear to have general validity?) and are the taxonomy and matrix useful in examining other technology development areas? In this chapter, then, the representativeness of the nuclear development programs is compared with analogous technology developments by drawing upon a limited number of other case studies that fit the criteria set out in Chapter 2.[6] This can help establish the utility of the analytic framework as well as lay the foundation for later model building and policy research on this topic. Furthermore, this exercise is more than merely intellectual, for "if we can discover empirically the policy conditions underlying our political patterns, we have a basis for better public policies."[7]

General Comparative Framework

The following comparison of government-sponsored, civilian technology development programs specifically adopts the perspective of multiple objectives and attributes posed in Chapter 2. The interactions between the objectives and the attributes are displayed in matrix form (see Figure 2-1) to represent the possible relationships. The matrix permits inquiry about

---

[5] Lowi, "Four Systems of Policy, Politics, and Choice," p. 299; emphasis in original.

[6] Or what Verba, "Some Dilemmas in Comparative Research," has called a "*disciplined* configurative approach," p. 117; emphasis in original.

[7] Lowi, "Four Systems of Policy, Politics, and Choice," p. 309.

which attributes of a technology development have the greatest influence in determining and carrying out a given public objective or set of objectives and about how different political objectives are manifested during the course of a development. The objectives column in the matrix is couched largely in terms of political-economic goals; the attributes are cast in terms of technology development. The matrix reflects the basic question considered here—namely, the government role in sponsoring advanced technology development and diffusion, and more generally, the economic and political forces that shape technology RD&D programs.

The frame of reference is strictly descriptive of what has been observed in the nuclear reactor programs and other civilian high technology development programs. We are not proposing a normative set of goals or what objective an organization should have pursued, nor are we prescribing what future institutions in a similar contingency ought to advance as their principal objectives in a technology development program. Instead, the objectives examined reflect what governments and their component organizations appeared to prescribe as policy goals, and the attributes are the variables that influenced the choice of those objectives.

Although each attribute can be perceived as affecting each objective, one can plausibly posit and, in some cases, demonstrate that different attributes are more important than others in determining the selection and possibly the successful attainment of the various objectives. This relative emphasis implies a number of policy-oriented hypotheses. In Figure 9-1, the hypothesized importance of certain attributes to a given objective is marked by crosses placed in the analytic matrix. These represent an initial attempt to identify which variables might be more critical in relation to the specific objectives.

Scientific Research for Science's Sake

If science for science's sake were a motivating national objective, one would expect to find an *active scientific community* that would be presenting evidence why a policy of government sup-

Objectives of Technology Development / Attributes of Technology Development	1. Active Scientific Community	2. Integrated Tech. Del. System	3. Concomitant Military Program	4. Early Starter	5. National R&D Heritage	6. Multiple Technology	7. Resources Invested
A. Science for Science's Sake	+		+		+		
B. National Prestige			+		+		+
C. Development of Efficient Technology		+				+	+
D. Political Equity	+	+				+	

Figure 9-1. Observed Interactions Between Objectives and Attributes

port for scientific and technical research would be a worthwhile governmental expenditure. In every nation examined, a group of scientists actively lobbied the government to build and finance basic scientific research facilities and train personnel for nuclear research. The construction and furnishing of such facilities was a first priority for government, even in West Germany after much of the fundamental nuclear research on atomic power plants had been completed. This attribute is not unique to nuclear and thermonuclear research; astronomy, high energy physics, and biological research programs all have active scientific constituencies pressing for government support.[8]

[8] See Lambright, *Governing Science and Technology*; and Greenberg, *The Politics of Pure Science*, for examples of this ongoing relationship.

The evidence from the nuclear reactor experiences suggests that the presence of a *concomitant military program* was influential in generating government support for the science for science's sake objective. The relationship in this study may have been because five of the six nations in the sample began their nuclear reactor research as an outgrowth of weapons research. The weapons or military research orientation was largely responsible for developing the national scientific stock in nuclear research, even if the personnel soon left weapons research for more civilian-oriented projects. The support that the Office of Naval Research (ONR) provided for universities to conduct research in the physical, medical, and biological sciences before the formation of the National Science Foundation underpins this relationship. It was hardly coincidental that Alan T. Waterman, the Chief Scientist for the ONR, became the first director of NSF.[9]

Third, the *national R&D heritage* attribute would support the science for science's sake objective by fostering the scientific and technical communities, which, in turn, would argue that government funds for their research were a worthwhile investment. Furthermore, they could point to the past benefits derived from the national R&D endeavors as proof of their claims. By and large, the evidence from the nuclear power reactor studies supported these assumptions; nations that manifested the science as a consumption good objective possessed a fairly robust R&D heritage. The scientific supporters of nuclear energy were quick to point out how continued government support was necessary to ensure the continuation of that heritage. However, it is impossible to validate or reject this hypothesis with any confidence from the present data base because all nations included were able to draw upon an industrial R&D heritage.

In summary, if the support of scientific research is a distinct objective in a technology development program, there is (1) an active scientific community encouraging and formulating as well as conducting the research; (2) a concomitant military program acting as an alternative source of funding, thereby supporting or at least complementing basic research efforts; and (3) a strong na-

---

[9] Greenberg, *The Politics of Pure Science*, ch. 7.

tional R&D heritage serving as both a base and a valuable precedent. The predominance of these attributes seems to indicate that a nation will pursue the objective of science for science's sake.

National Prestige

If one posits national prestige as an objective underlying a technology development, one might expect to find the *resources invested* to be the most critical attribute defining the program. These represent a national commitment to a specified goal and the sunk costs that the institutional actors are unwilling to write off or abandon. For national prestige, one would expect to find the government to nongovernment ratio more biased toward the former because the preservation of national prestige is more likely to be perceived as a governmental function (i.e., there are fewer private returns to be gained).

The American decision to send a man to the moon was due less to the development of American technology than to the Soviet Union's orbit of the first earth satellite and the embarrassment of the Bay of Pigs.[10] When Almond polled European and American public opinion, he found an "extraordinarily high" public recognition of initial space accomplishments, largely because the United States had already staked its national prestige on its space program. To this end, it had spent considerable resources and, in so doing, indirectly raised its citizens' collective space consciousness. Therefore, it was not surprising that "almost every respondent in the countries surveyed was aware of the launching of the first satellite and of the fact that Russia had launched it. Subsequent awareness of the first American success was also high."[11] Almond concluded that these respondents also perceived space accomplishments as integrally linked to national strength:

> One of the most stable popular beliefs of the postwar era—the belief in the scientific and technological superiority of the United States—has

[10] See Logsdon, *The Decision to Go to the Moon.*

[11] Gabriel A. Almond, "Public Opinion and the Development of Space Technology," in Joseph M. Goldsden, ed., *Outer Space in World Politics* (New York: Praeger, 1963), pp. 73-74.

been rudely shaken, and its place had been taken by anxious estimating that fluctuates with each report of a significant step forward in satellite launchings.[12]

Kennedy, recognizing and responding to the "debt" incurred and to the public perceptions, made the reestablishment of U.S. international prestige the major objective of the Apollo program and was willing to devote considerable resources to attaining that goal. For the Apollo decision, these additional resources meant an increase in the NASA budget and personnel from $1.8 billion and 60,000 people in 1961 to $5.2 billion and 420,000 people five years later at the height of the program.[13]

By the same token, the U.S. SST, which was viewed by many as little more than an overt attempt to maintain U.S. prestige in commercial aviation, was primarily funded by the federal government. Had the American SST been less of a prestige development (i.e., promised more likely commercial applications and financial return), one would have expected to find private industry contributing more to its design and development than the less than 5 percent of the total costs that Boeing funded.[14] One resulting hypothesis is that nongovernmental sources are less likely to contribute to the development of a technology whose acquisition is viewed as a matter of national prestige.

This relationship is certainly supported by the nuclear reactor development data. As nuclear power reactors became symbols of national prominence, nations were willing to invest large amounts of money to develop their own national industry and the accompanying prestige high technology bestows. Even the West German government and electrical construction industry, which decided to "buy in" the reactor technology, went to expensive lengths to develop a domestic capability. Additional support is provided by the reluctance of the British government to relinquish its nuclear heritage by treating its technically deficient reactor program as an economic sunk cost for fear that the pro-

---

[12] Ibid., p. 77.

[13] Cited in Schulman, "Nonincremental Policy Making," p. 1355.

[14] Eads and Nelson, "Government Support of Advanced Civilian Technology," cite the high financial costs and risks borne by IBM in developing a new computer generation as a counter example to government-funded technology developments.

gram's demise would cast aspersions on the nation's prestige. Finally, the ratio of government to nongovernment spending on the development of nuclear power reactors was generally much higher when the national prestige objective was high—that is, the development period.

A second critical attribute of the national prestige objective for the nuclear reactor programs was the presence of a *concomitant military program.* Nuclear weapons were initially sought by each of the nations studied (except, of course, West Germany) as either a wartime weapons development or an indication of its great power status; in four out of six cases, nuclear weapons can be treated as symbols of national prestige. The technical carryover from atomic weapons research to nuclear power reactors and the prestige ascribed to the former implies that a certain amount of prestige value was accorded to nuclear power reactors in the wake of nuclear weapons systems; reactors served to reinforce the great power status conferred by nuclear weapons.

The third pivotal attribute that characterized the national prestige objective in these cases was the *national R&D heritage.* The underlying hypothesis is that the presence of an honored tradition in R&D and technology innovation is one that a nation's decisionmakers and its industries wished to enhance or, at worst, not permit to be denigrated. French and British spokesmen both cited their earlier national contributions to civil aviation as an important—although partial—basis for their investment in supersonic transportation. One of the reasons underlying Kennedy's Communications Satellite Act of 1962 was to protect the long-standing tradition of American leadership in communications technology.[15] German scientists argued that a nuclear research program was important to reestablish German R&D credentials as an advanced industrial nation and, as a matter of national pride, to emphasize her industrial and technological recovery.[16]

The relative influence of another attribute, *active scientific*

---

[15] See Jonathan Galloway, *The Politics and Technology of Satellite Communications* (Lexington, Mass.: Lexington Books, 1972); and Judith Tegger Kildow, *INTELSAT: Policy-Maker's Dilemma* (Lexington, Mass.: Lexington Books, 1972).

[16] Keck, "West Germany Science Policy Since the 1960's."

community, is thought by many to contribute strongly in the pursuit and achievement of the national prestige objective. Perhaps more than any other interest group, the scientific and technical communities would suffer the greatest loss and embarrassment were a nation's scientific prestige perceived as degenerating. Such a perception might result in an emigration of the nation's scientific stock to other nations with demonstrated technological excellence in search of peer and financial support. Other interested groups, such as private industrialists, might be more willing to accept a decline in national prestige as presaged by a declining rate of technology developments because there is ample evidence that the technological innovators are not always commercially the most successful purveyors of that technology.[17] In fact, Gilpin argues that the Japanese economy is deliberately predicated on "buying in" on proven technologies and then realizing profits on superior manufacturing and merchandising techniques.[18] But this cautious, contemplative attitude is less attractive to the scientific communities, which ascribe a closer connection between the concurrent declines in national technology and prestige and thus politick strenuously against the loss in prestige.

The evidence drawn from the nuclear reactor development cases does not support or reject this posited relationship with any confidence. Certainly the scientific community was instrumental during the early development days of nuclear energy when the prestige objective was probably at its zenith. However, it is not clear whether the considerations of the scientific community were more attendant on science for science's sake, which was also a motivating objective during that same period. During the dissemination period, when national prestige objectives began to reassert themselves, the scientific community did not play an instrumental role. Thus, the evidence seems to argue against this proposition. In any case, this attribute was certainly

---

[17] There is an impressive literature on this subject; for two excellent overviews, see F.M. Scherer, *Industrial Market Structure and Economic Performance* (Chicago: Rand McNally, 1970), ch. 15, "Market Structure and Technological Innovation;" or James M. Utterback, "Innovation in Industry and the Diffusion of Technology," *Science* 183, 4125 (February 15, 1974):620-626.

[18] Gilpin, "Technological Strategies and National Purpose."

less important in motivating the national prestige objective than the previous three.

To review the national prestige objective briefly, it was most likely defined by the presence of (1) government-invested resources, (2) a concomitant military program, and (3) a strong national R&D heritage.

Development of Efficient Technology

The development of an efficient technology is seen to encompass the development and civilian application of a new technology in such a manner that it is economically competitive with other technologies serving the same function (e.g., an SST and a Boeing 707). A central attribute for achieving this objective seems to have been an *integrated technology delivery system* (TDS). (The converse is not necessarily true. An integrated TDS does not guarantee the efficient development and application of a technology; the U.S. SST was not terminated because of an ineffective TDS.)

The critical nature of this relationship was borne out by the six national nuclear reactor development and dissemination experiences. The United States, West Germany, Canada, and (from the evidence available) the Soviet Union all displayed an integrated TDS. As seen above, the British and French undermined their national TDS for most of the period studied by treating the vendors strictly as suppliers and the consumers as a captive market, denying them a role in the development and dissemination processes. Indeed, it was not until the French TDS actors reached a state of equilibrium similar to that of the United States and Germany that the French government was willing to move to a more efficient reactor system, in spite of its development outside France.

This finding is supported by other cases of technology development. The French and British governments both treated their national airlines as captive consumers and gave them little opportunity to participate in the Concorde deliberations, thus short-circuiting the concept of an integrated TDS.[19] The inability or

[19] Witness, for instance, the pressure brought against BOAC Chairman Sir Giles

refusal of the federal government and the aerospace vendors to override consumer (municipalities) objections has prevented the widespread dissemination of an inefficient personal rapid transit technology. Perhaps one of the reasons for the excellence of the American and West European medical research delivery system is that the government/vendor/consumer relationship has been ironed out over a number of years and iterations. In the PRT and medical cases, a fairly integrated TDS was influential in determining the diffusion of effective technologies while, in the Concorde, the diffusion of an ineffective technology can be partly attributed to the truncated TDS.

The second variable motivating the development of an efficient technology is the influence and level of *resources invested*. The history and amount of resources invested create institutional and political pressures to continue the research through traumas of cost overruns that appear so endemic to advanced technology developments during the demonstration and dissemination stages. There is a great deal of evidence that the expensive portion of technology RD&D is in the demonstration of its civilian application—that is, its economic competitiveness with alternative technologies.[20] Lacking the earlier commitment generated by the past resource investment decisions, the program might not be able to muster sufficient support to warrant continued development. This assessment obviously presents some definitional problems—specifically, what level or amount of prior resources invested is necessary for decisionmakers to appreciate the influence of the earlier funding?

The national nuclear reactor development experience supports the observation that development and diffusion stages are the most expensive stages of technology RD&D but adds little precision as to what levels of resources invested might be necessary to

---

Guthrie when, in late 1963, he announced that he wanted to cancel BOAC's order for the British-built VC-10 and replace them with Boeing 707s. His action was publicly disavowed by the minister of aviation who, in a speech before Parliament, made it clear that "The terms on which BOAC finally make up their minds must be a matter for their commercial judgment of course . . . [but] in consultation with me." Quoted in Costello and Hughes, *The Concorde Conspiracy*, p. 68.

[20] See Scherer, *Industrial Market Structure*, ch. 15.

maintain organization support for questionable programs. As the West German example demonstrates, there seemed to be no way of inexpensively developing a domestic reactor program; the German program was experiencing serious growing pains until the federal government began to offer larger subsidies and outright grants to encourage the development of the nuclear industry and the proliferation of nuclear facilities. According to Burn, the British AEA spent as many resources on its reactor development program as the U.S. AEC, but the results differed markedly.[21] The amount of resources required before they are perceived as cause for continued support obviously varies from program to program. It is almost certainly too important to be left unconsidered for this objective, but for the time being, the threshold level or standard (either in absolute or relative amounts) must be left unspecified.

Other developments offer additional evidence but precious little light. Clearly, large amounts of money were spent on the technology developments considered here but just "how much is enough" is a complex, perhaps unanswerable, question. However, significant amounts of R&D monies have been invested in technology developments whose "success" in reaching an efficient technology is debatable. Costello and Hughes estimate the American SST cost the taxpayer over $880 million (plus an additional $1.5 million in termination costs).[22] The Concorde development cost close to £1,100 million and lost over $54 million in its first full year of operation.[23] BART cost close to $1.6 billion and operated in fiscal year 1976 with a $40.3 million deficit.[24] Obviously, these figures represent significant amounts of resources invested, and yet there is little correlation between the investment and the development of an efficient technology as defined.

The third key attribute to the efficient technology objective is postulated to be the presence of *multiple technology options.*

---

[21] Burn, *The Political Economy of Nuclear Energy*, pp. 84-85; an AEA spokesman's comment on Burn's relative assessments in Walsh, "Nuclear Energy: New Study Assails British Program," pp. 1350-1352.

[22] Costello and Hughes, *The Concorde Conspiracy*, p. 285.

[23] The *Los Angeles Times* (January 22, 1977), Sec. I, p. 4.

[24] Webber, "The BART Experience," pp. 92, 103. The experience to date of the recently opened Washington, D.C., system scarcely seems to hold much greater promise.

Although multiple technology alternatives do not assure the production of an efficient technology (witness the number of different rapid transit systems), a choice among alternatives is one of the most effective ways for bringing a technology into commercial application. As noted above, this relationship was positive in military weapons systems acquisition strategies and it can be presented as equally applicable in the RD&D for civilian technologies, such as water desalination (the vertical tube evaporator and the multistage flash distillation processes), hydraulic knee prosthetic devices, and INTELSAT communications satellites (until recently, the Hughes and TRW designs).[25]

The *multiple options* attribute appears to be important because it allows a larger leeway for technical error and a hedge against technical uncertainties. The U.S. reactor systems that were initially thought to be the most promising were later superseded by other reactor technologies whose reliability and lower costs in lieu of technical sophistication and theoretical superiority proved superior for delivering nuclear-generated electricity. Likewise, the Soviet Union's multiple reactor development program gave Soviet leaders a wide choice, even permitting them to discard a reactor design that was to become a principal American design (the BWR). Thus, a number of alternative technical designs allowed the decisionmakers to choose the more efficient options. Still, it is impossible to demonstrate this attribute's influence conclusively. The Canadian program built only one reactor system yet was successful in achieving this objective, and none of the many PRT designs has passed the market efficiency tests.

Finally, some have argued that a *concomitant military program* is critical in achieving an efficient civilian technology, an assertion that the nuclear reactor evidence does not support. The United States and the Soviet Union had ongoing weapons programs and achieved an efficient technology; France and Great Britain had weapons programs but failed to reach that goal; West Germany and Canada had no weapons programs, but both attained some measure of success in developing and disseminating a reactor system. The SST programs similarly fail to support this

---

[25] For descriptions of the first two, see Baer et al., *Analysis of Federally Funded Demonstration Projects: Supporting Case Studies.*

hypothesis. The Concorde design teams overcame major technical problems without the benefit of military bomber research, and their American counterparts were forced to make significant design modifications even though they had the B-70 research as a starting point.

If the successful development of an efficient technology as an export commodity were a motivating objective, one would expect to find the same attributes observed but with perhaps slightly different priorities. For example, the existence of an integrated technology delivery system would continue to be critical. Nations that experienced a minimum of institutional strife during the development of their reactor systems had the greatest success in exporting their nation's atomic reactors. The British reactor program, which was more characterized by the insulation of its TDS actors from each other than any other nation examined, was able to sell only two reactors abroad although the English AEA made perhaps the most concerted export effort. Similarly, the French program kept its TDS actors at arm's length from one another until the late 1960s and was unable to place any reactor export orders, even with the intense political pressures France placed on Germany to buy the French gas-graphite reactor. On the other hand, Canada had few institutional problems and enjoyed some success exporting its heavy water reactor, which, like the British and French models, consumed natural uranium. This same degree of institutional harmony and its relationship to exporting a commodity can be seen in the case of the British-French development of the Concorde as an exportable commodity. The fact that BOAC and Air France were captive markets isolated the developers and vendors from consumer requirements, a failure that ultimately proved decisive when the SST proponents tried to export the airplane to markets their governments could not control.[26]

One can posit that governmental *resources invested* in developing a given technology for export purposes is a critical attribute because they commit the public sector to develop the technologies as a means of improving or even correcting national

---

[26] See Costello and Hughes, *The Concorde Conspiracy*; and Wilson, *The Concorde Fiasco*.

balance or payment problems. Nongovernment organizations could view the technology as a likely means for entering new markets and therefore invest significant amounts of their own resources into the development. Government and nongovernment investments help bind the participants to their respective (sometimes coincident) objectives. This behavior was clearly evident for nuclear power reactors. All of the major government actors in each of the programs studied (with the possible exception of the Soviet Union where the details are unclear) devoted considerable resources to developing an exportable reactor. The U.S. LWR was the most difficult to export because of its requirements for enriched uranium, but the 1954 Atoms for Peace proposal making U-235 available to nations that did not have the technology or resources to fabricate enrichment facilities overcame that obstacle.

Similar behavior was also apparent for the Concorde and American SSTs. The British and French governments both viewed the Concorde as a means of recapturing the civil aviation market from the United States, a view shared by the British Aircraft Corporation and Sud Aviation and, coincidentally, the American aircraft industry. The various national actors were prepared to contribute large amounts of their resources to the development of a technology that was largely dedicated to the external market. According to Costello and Hughes, the U.S. government spent over $828 million on the American SST before canceling it; Boeing and other private concerns spent an additional $52.3 million of their own funds.[27]

Upon close examination, the resources invested attribute requires a slightly broader definition, especially in the nuclear reactor context. For export reactors, different national governments competed on more than strictly technical bases. Heavy subsidies underpinned most foreign reactor sales,[28] technological "side payments" were offered to enhance a nation's offer,[29] and, in

---

[27] Costello and Hughes, *The Concorde Conspiracy*, p. 285.

[28] In retrospect, the subsidization policy can be seen as contradicting most governments' nonproliferation policies. This is especially clear in the case of the United States; see Wohlstetter, "Spreading the Bomb Without Quite Breaking the Rules."

[29] The most obvious example is the German willingness to include a nuclear

some cases, large "commissions" were given in recognition of the selection of a particular nation's reactor system.[30] In this expanded sense, then, the term resources invested covers a wider set of expenditures than were originally defined. Had the original definition held—that is, had resource invested consisted only of the national R&D costs (both public and private)—then that attribute might be less important. But under the broader definition, it is safe to assume that resources invested is second only to the integrated TDS as a critical variable explaining the pursuit of developing an efficient technology for export purposes.

Finally, nations that developed multiple technology options enjoyed greater export success than those that did not. This is consonant with the school of thought that multiple technology approaches generally provide the greatest benefits in technology development. The West German and American successes in exporting their reactors stand in contrast to the French and British failures. However, once again, the evidence of this relationship is somewhat mixed. The single Canadian reactor had a limited amount of success in the export market, and the multiple Soviet reactor development program has produced only one reactor sold outside the Soviet bloc. In both the domestic and external markets, one might justifiably suspect that this particular attribute has only limited explanatory power compared with the previous two.

---

reprocessing facility in the nuclear reactor package the West German KWU sold to Brazil; see David F. Belnat, "Brazil's Nuclear Program Aimed at Filling Energy Gap," *Los Angeles Times* (May 30, 1977), pp. 1, 6; the Brazilian program is reviewed in the political context of Latin America by Karen DeYoung, "Latin Americans Hurry to Catch up in Nuclear Power," *Washington Post* (July 8, 1977), pp. D1, D5; also see William W. Lowranc, "Nuclear Futures for Sale: To Brazil from West Germany," *International Security* 1, 2 (Fall 1976):147-166.

30 Even Canada, which rarely stressed the export potential of its CANDU reactor, seems to have engaged in this activity; John Foster, the President of the AECL, was dismissed from his post, in part "over the payment of $10.4 million in commission for outside assistance in the sale of nuclear reactors to South Korea and Argentina, which the Auditor General said had not been supported by satisfactory explanations of the services for which the money had been paid." Numbered Swiss bank accounts and Israeli-based agents were also cited. Robert Trumball, "Canadian Minister Dismisses Chief of Atomic Agency," *New York Times* (July 8, 1977), pp. D1, D5.

Two other attributes warrant discussion because many advance their importance in developing an exportable technology. The first, *national R&D heritage*, proposes that nations and national industries with an important R&D tradition have generally been accustomed to predicating a positive balance of payments posture on the export of the products or processes of their advanced, innovative technologies. Therefore, this attribute would point to the development of an efficient technology with an explicit export objective in mind. The British program, perhaps more than any other nation's, personified this position. England had been a great manufacturing and exporting nation during her empire years and there was little reason for the British government and industries to think that the postwar situation would be any different except for a change in the technologies.[31] Harman documents American computer R&D and how American computer exports based on that research heritage dominated the foreign marketplace;[32] computer technology has been perceived as a significant element in the widely debated technology gap between the United States and the other OECD nations.[33] This work is unable to shed additional light on this particular proposition. All of the nations involved could point to the products of their R&D establishments, and yet some had greater success in exporting reactors than others. Indeed, the British and French could both claim to have a more extensive R&D establishment and a richer heritage than the Canadians, and yet the CANDU system was more successful than its French and British competitors.

The *early starter* attribute is often hypothesized to be important to exporting a technology, holding that one of the surest ways to capture the export market is to be the first to demonstrate the commercial applications of a given technology,

---

31 In September 1962, British Prime Minister Macmillan stated that "Britain should make not just things which anybody can learn to make, but the difficult things which need precision and highly skilled workmanship . . . : nuclear power stations, computers, supersonic aircraft." Quoted in Costello and Hughes, *The Concorde Conspiracy*, p. 55.

32 Alvin J. Harman, *The International Computer Industry: Innovation and Comparative Advantage* (Cambridge, Mass.: Harvard University Press, 1971), especially "The Computer Industry in an International Setting," ch. 2.

33 OECD, *Gaps in Technology*, pp. 217-220.

which implies an early start in its development. There is substantial evidence that the first examples of technological innovation enjoy a significant advantage in the international market, probably supported by the perception that the monopoly of a weapons system provides a significant military advantage. Experience in computers, electronics, and metallurgical processes all underpin this hypothesis.[34] This was a driving motivation in the development of the Concorde, especially during its initial planning by British engineers and government officials. It was critical, British aeronautical engineers explicitly argued, that the British SST be available before the U.S. aerospace industry could produce an American SST.

The evidence from the nuclear reactor development programs does not support this putative relationship. Britain, which had the first operating electricity-generating reactor in the world and early Magnox reactor sales to Italy and Japan, was left far behind in the reactor export competition by the United States and, most tellingly, by the West Germans who were as late starting as the British were early. The cases suggest that being an early starter may, in fact, undermine the subsequent marketing of a technology if by being an early starter one becomes committed to an outmoded technology. The British adherence to the gas-graphite systems in the face of light water technology recreated the situation in civilian aviation where the English Comet preceded the Boeing 707 by five years yet, in the end, succumbed to the newer American technology, thus surrendering a lead in civilian aviation Britain has unsuccessfully tried to reclaim ever since.

In summary, the development of an efficient technology for either domestic or export purposes as an objective motivating a technology development is characterized by (1) an integrated technology delivery system, (2) the resources invested by both government and nongovernment sources, and to a lesser extent, (3) the development of multiple technological alternatives.

Political Equity

The political equity objective was included as a manifestation of political interest group theory, although the attributes set forth

[34] Ibid.

in the analytic framework as defining this objective are not those found in traiditional political group literature. The attributes posed here are surrogates of the different interest groups as they would make their claims upon the government for support or to cause the government to engage in exercises we define as political equity.

The attribute that most reflected this objective was *multiple technology options* because it represents different organizations seeking government support for their respective activities or, for nuclear reactors, different designs. The process, then, of allocating resources to the competing groups and their approaches can be treated as part of the political equity objective. This can be seen in many civilian technology developments. The two American SST designs (by Lockheed and Boeing) represented more than two technical solutions to the same set of problems; they also represented two different associations of organizations and their supporters, each maneuvering for government support. Even in the centralized Soviet decisionmaking process, the accepted explanation for the multiple technology approaches that characterized the Soviet R&D strategy in reactor and other developments is the existence of a number of different groups, each striving for support, and the government engaging in an equitable distribution of funds and personnel (obviously within certain defined limits) to the contending factions. In the United States, AEC, reacting to the vendor and user demands, encouraged as wide a base of industrial and utility participation as possible, even sometimes contracting groups whose assets and organization were not suitable for the task.[35] Government subsidies and grants were distributed among these different groups whose existence reflected a number of different approaches to the problem of nuclear power reactors.

A second attribute in defining this objective might be the presence of the *integrated TDS*, which represented the three major institutional actors in the RD&D and dissemination processes. At times the actors were cooperative (or perhaps "coopted"), and the transitions between them were without incident; at other times, they were fractious. In all cases, the TDS

---

[35] See Allen, *Nuclear Reacters for Generating Electricity*, p. 74.

concept demonstrated the demand of the different actors upon the political system. The German case is particularly interesting in that the vendors and consumers urged the central government to assume a more active role, even if it meant a constriction on their latitude. The reason, of course, is that increased government participation meant more resources, so that although the vendors' and users' freedom might be reduced, they would have better access to more funds.

Finally, the *active scientific community* was defined as composed of scientists and technicians who were largely dependent upon government support for the development of advanced technologies and more important, their professional livelihoods. It is thus a textbook example of a political interest group, one that might be expected to voice its needs through the political processes. The data from the national nuclear development and many other scientific research programs underpin this assumption. The active scientific community attribute seemed to be a valuable indicator of the political equity process; the scientific community was—and is—a lively, articulate, and highly visible interest group that consistently demanded and obtained a sizable government contribution to both applied and pure scientific research.[36]

Resources invested have not been posited as a critical attribute of political equity because the allocation process entails a distribution of these assets. The reason for its exclusion is that political equity, as defined, is somewhat less concerned with the size of the original investment than with the way the investment is distributed. Therefore, there is less attention to the amount of resources invested in this objective than in others.

In brief, the political equity objective is most cogently described by the (1) multiple technology options, (2) the integrated technology delivery system, and (3) the active scientific communities attributes.

---

[36] The impressive achievements of the scientific community as a political interest group are documented in Lambright, *Governing Science and Technology*; and Greenberg, *The Politics of Pure Science*.

Review

The relationships between the objectives and attributes observed in the national nuclear reactor development programs were critical in molding and directing the programs. Evidence that these relationships are not unique to these programs is provided by similar patterns of objective-attribute relations reflecting interorganizational tensions in other civilian technology developments programs. Taken either singularly or in combination, the relationships offer the basis for a testable set of policy hypotheses that point in the direction of a general model of government-sponsored civilian technology developments.

Toward a General Model

The general analytic framework for the development and dissemination of government-sponsored civilian technology presented above has been seen as somewhat useful in explaining past developments and offering policy observations applicable for future development. However, this does not mean that a derived model could rest on whatever limited laurels its ancestor might claim. A great deal of additional work would be required before the analytic framework could be considered a general model and a policy tool for those interested in understanding the interaction of technology and social forces and to policymakers in technology areas. This work falls into three broad efforts: (1) detailing the relationship between the objectives and the attributes; (2) increasing the precision of the model; and (3) model validation and respecification. Each, of course, is highly dependent on the other and would necessarily set in unison. For the immediate purposes, however, they are treated separately. This section offers some suggestions in addressing these lines of research.

Interrelationships

Before the analytic framework can assume the trappings and responsibilities of a general model, one must begin to delineate with some clarity the relationships among the various objectives and attributes. The reason is that a model must offer a set of logical, *interconnected* statements that explained the phenomena within its purview. The analytic framework does not, as yet, do this, and a model derived directly from it would distort the realities of public policy decisions. Policy decisions, of course, are predominantly characterized by tradeoffs among contending objectives. For a model to depict the policy arena with any verisimilitude, it must reflect and include those interrelationships.

As the analytic matrix is currently framed, the individual objectives are treated as if they were mutually exclusive. Different attributes describe the objectives with little attention paid as to how each might influence the others. In the actual decisionmaking processes, this is not the case. A nation's policymakers might well downplay attaining the efficient technology objective in favor of the national prestige or policy equity objectives, as was the case of the British nuclear reactor's systems and the Anglo-French Concorde. A similar problem exists with the attributes; for instance, the relationship and tradeoffs between the active scientific community and the concomitant military program or the national R&D heritage is certainly pertinent but it has not been described in either generic or specific terms in this book.

In short, before we can talk about achieving a general model using the categories utilized above, we must delineate the relationships among all the component parts of the proposed model.

Model Precision

A second task would be to add greater precision and clarity to the model and its components by moving, however modestly, in a more empirical direction so that the policy propositions generated by the objective/attribute matrix could be more rigorously posed and tested. The definitions used for the analytic framework were meant to serve only as first approximations of the RD&D and dissemination processes and are hardly definitive. Future specification of the model should make the definitions and the supporting data less subjective and more empirical in their content. One example is the national R&D heritage attribute. Recent work by the OECD has tried to formulate standardized cross-cultural measures of R&D and its influence in a nation's economy;[37] others have attempted to delineate the effect of the R&D constituencies and their contributions to the nation.[38] Efforts to sharpen the framework's definitions should attempt to incorporate these new indices, however tentative they might be at present.

The definitions of the objectives and attributes should move toward some type of quantification. The unavailability of appropriate data will inhibit a crucial part of these efforts. A more precise definition will not overcome the data availability problems in cleanly measuring the national prestige objective (see Appendix A). A great many amorphous elements are hidden within that objective and others and make the strict quantification characteristic of the behavioral sciences highly unlikely in this area. Yet without such attempts, additional testing of the policy propositions would be little more than impressionistic. This is certainly not a call to reduce the matrix and propositions to exercises in regression analysis. The point is not to demand empirical purity but to move away from strictly subjective evaluation. In this way, the intellectual and policy merits of the model can be more objectively tested and deliberated.

If additional objectives and characteristics increased the explanatory and predictive capabilities of the model, they would increase the model's usefulness, both as an intellectual tool for explaining past developments and as a policy analytic framework. Two possible candidates for new attributes might be "national culture" and "time of development"; a new objective might be "create new industry" or to isolate a "technology for export" objective. Naturally, one must be careful not to add too many new variables and objectives, lest the model become unwieldy. But certainly one should scrutinize the technological and political environs of other developments and other nations to ascertain if

---

[37] *The Measurement of Scientific and Technical Activities* (The "Frascati Manual") (Paris: Organisation for Economic Co-operation and Development, 1976), is a recent example.

[38] See the essays in T. Dixon Long and Christopher Wright, eds., *Science Policies of Industrial Nations* (New York: Praeger, 1975).

there are any additional development and dissemination phenomena not currently represented and to ask what other objectives and attributes might be amended to capture them.

A final area for possible model building is less concerned with increasing its precision than with expanding its scope. As currently defined, the model considers only of large-scale, government-sponsored developments of technology destined for the civilian sector. Implicitly, it deals solely with highly industrialized societies. It would improve the worth of the proposed model if some of these limitations could be relaxed. For example, a sizable case study literature already exists concerning lower cost government development of civilian technology, which, if it could be melded into the model, would expand the purview beyond high cost systems.[39] It might be possible to increase the model's scope so that high cost developments that were largely funded by the private sector could be included; IBM's development and commercialization of the Model 360 computer system was an expensive undertaking financed entirely within the private sector (ignoring, for the moment, government tax incentives for private R&D).[40] The consideration of technology developments in the less industrialized nations would permit the model to consider a new provocative set of policy issues, those relevant to the development of an entire industry or even an economy rather than a single technology. Finally, the obvious expansion of the model to encompass military technology developments presents serious conceptual problems. For instance, in terms of the TDS for military technologies, the developer is the consumer and dominates the vendor; that is, the distinct roles of the three are greatly obscured. But this does not render the concept irrelevant. One could plausibly define one branch of a service as the developer and another as the consumer, both proceeding on different assumptions and after different objectives. It is one thing to declare the problems are difficult; in this case, to continue by declaring that they are impossible is not only a *non sequitur*, but not necessarily true.

---

39 See Baer et al., *Analysis of Federally Funded Demonstration Projects: Supporting Case Studies*, for a number of excellent case studies along this line.

40 Eads and Nelson, "Government Support of Advanced Civilian Technology."

Statements of the model could follow any of the avenues of research outlined above and generally improve both the precision and utility of the model. However, none of these avenues can be traveled without the acquisition of new data and the promise of careful testing.

Model Validation

To test the model, additional data must be accumulated beyond those studied here. The exact data collected depends, of course, on how the model is specified (what are the objectives and attributes and what is the scope), but for convenience's sake, we can group the new evidence requirements into two categories, additional case histories of nuclear power reactor dissemination and totally new examples of different technology developments.

For the immediate purposes of posing energy policy questions for the analytic construct, additional national case studies of the development and dissemination of nuclear power reactors would be most relevant. The focus on industrialized nations could continue; to a large extent, Sweden has developed its own power reactor system, Japan has built a major reactor industry based upon foreign licensing (in much the same way as West Germany), and Switzerland has a higher percentage of nuclear-generated electricity than any other nation.

It would be perhaps even more interesting to expand the original scope of the analysis and examine reactor programs in nations that are not as industrially advanced as the Western European nations. The two immediate candidates are India and Israel.[41] Finally, nations that are currently preparing to engage in large nuclear reactor programs—Spain, Iran, Argentina, and Brazil—would provide additional evidence of the model's utility, especially in regard to its policy guidance capabilities. The inclusion of these new cases would certainly enhance the explanatory powers of the framework and possibly shed light on some of the more ambiguous variables, for instance, the national R&D

---

[41] For a recent example, see Onkar Marweh, "India's Nuclear and Space Program: Intent and Policy," *International Security* 2, 2 (Fall 1977):96-121.

heritage, because some of the new nations would lack this particular attribute.

If the model is applicable beyond the development of nuclear or energy technologies, its scope must be expanded to include other technology developments that fit within its boundaries. These could be examples of a number of technology developments in one nation or, preferably, technologies viewed across a range of nations. For example, industries in the United States, Japan, and the western European nations have all worked to develop new forms of high speed rapid transit. Other forms of energy research could be included, such as fusion, magneto-hydro-dynamics, and breeder reactor development programs. A broader base of technologies would permit greater discrimination among the attributes and objectives because greater variation would probably be observed.

The acquisition of new evidence must, of course, be tempered by the accessibility of the required data. Serious definitional problems restrict the researcher's ability to gather some kinds of data. The necessary data might not exist as defined. This is true for the present model (see Appendix A) and probably would continue for future iterations, especially if a larger number of disparate nations is surveyed. For example, the industrialized nations are beginning to record various indices of R&D activities and economic growth, but these measures are much less available in the underdeveloped nations. The accessibility of data presents other problems, such as the closed nature of the Soviet system of government; inferential license should only go so far before being revoked.

Summary

Even though the preceding discussion distinguished between the model delineation and validation efforts, the two are intimately joined in methodological fact. The specification of the model or its expansion without a thorough appreciation of the existence and accessibility of the necessary data would have a low probability of success and be little more than an academic exercise. Similarly, the pell-mell collection of data on new countries and technologies without the theoretical and analytic structure could easily leave the researcher with evidence that is irrelevant to the model's needs. Neither situation is desired, nor is it inevitable so long as the special requirements of model definition, data acquisition, and the interaction between them are closely considered.

As proposed, the general model of technology development and dissemination promises a number of intellectual and policy insights. It permits the aggregation and synthesis of a large amount of cross-national data in such a manner that patterns can be discerned; conceptual and data deficiencies may be identified. However useful the framework might currently be, the potential benefits that might accrue from the suggested refinements are much more impressive. It is hoped that this work has stimulated researchers about the efficacy of the approach utilized so that others will be encouraged to continue where these efforts ended. Obviously, much remains to be done but even more remains to be gained.

Appendix A: Possible Measures for Model Objectives and Attributes

In rigorous physical or behavioral science models, the different components or variables can be subjected to precise definition, quantification, and measurement. The suggested model of government-sponsored civilian technology development would have greater utility if each of the objectives and attributes were similarly suitable to these precisions, especially for objective and attribute measurement. However, for such a difficult issue as the cross-national comparison of technology developments and at this stage of model construction, we cannot realistically hope to bring such exactitudes to bear. This shortfall should not be particularly unexpected. As Kaplan cautions, "In many areas of human behavior, our knowledge is on the level of folk wisdom; incorporating it in a model does not automatically give such knowledge scientific status. The maturity of our ideas is usually a matter of slow growth which cannot be forced."[1]

---

[1] Kaplan, *Conduct of Inquiry*, p. 279.

Chapter 2 devoted a great deal of attention to the definitions used in the analytic framework without distinctly discussing the appropriate measures and the availability of the data (especially quantitative) that compose the objectives and attributes. The proposed model's data requirements reflect Kaplan's concern that models of extremely complex situations "are often improperly exact: they call for measures that we cannot in fact obtain, or that we would not know how to use if we did obtain them."[2] Recognition of the situation does not, however, imply a stoic acceptance of the data definition and availability problems as a continuing condition. Model building is an ongoing process practiced by the person who first set forth the model and by those that follow, test, and amend it.[3] Therefore, it is a worthwhile exercise to suggest possible measures for the proferred objectives and attributes in the expectation that later studies will be able to build upon the current efforts by bringing greater precision to the model specifications and obtaining better data.

This appendix focuses upon data deficiencies. The definitions of the objectives and attributes discussed in Chapter 2 need not be repeated.

Objectives

Scientific Research for Science's Sake

This objective might be approximated by the amount of government money spent on areas of technology development within the type of technology or research area (such as high energy nuclear physics, biological research, or space research) over a given time interval. The importance the government places on this objective could be emphasized by looking at these budgets relative to the overall government scientific research or R&D budgets, or perhaps the research funds per scientist in a given area of

---

[2] Ibid., p. 293.

[3] This process is detailed at great length by Thomas S. Kuhn, *The Structure of the Scientific Revolutions* (Chicago: University of Chicago Press, 1970).

endeavor.[4] In the United States, this would include an examination of the NSF, NIH, and possibly the NASA and Department of Defense budgets.[5] What we would hope to show is the degree to which governments are willing to support scientific research with the tacit objective of underwriting the scientific community, even if its direct benefits to the population are somewhat ambiguous.

National Prestige

Demonstrating the presence, let alone the effect, of domestic and international prestige as an objective to a technology development program is difficult. In many cases, there are explicit statements about this objective by national decisionmakers, but questions of credibility must be raised. In other words, are appeals for technological patriotism more rhetoric than real government policy? A careful examination of government statements can be used, perhaps in conjunction with supporting evidence; content analysis might be a useful methodology. Alternative measures might be used to infer the prestige objective. For instance, are there reasons why a nation might wish to bolster its international prestige and hence be willing to promote a major technology development as a means to that end? A nation that has suffered a series of international setbacks (such as the loss of a war or an empire), that is attempting to assert its position in the international arena, or that would publicize a technology development as an indication of the superiority of a political or economic system might be considered as a likely candidate for the technology for prestige objective. An estimation of the prestige objective depends upon indirect criteria and inferential evidence, perhaps more than any other.

---

[4] For one possible example, see OECD, *Changing Priorities for Government R&D*.

[5] Greenberg, *The Politics of Pure Science*, emphasizes the roles of these agencies in the support of basic scientific research.

This objective will vary from technology to technology, but its general intent is to provide a measure of the ability of a nation's RD&D program to deliver a new technology able to compete on a price basis with the other technology alternatives available to that nation for fulfilling a specified demand. The problem of government subsidies interferes with a clean interpretation of whatever data on economic viability one might use. For example, a federal government grant might subsidize the capital construction costs of a PRT system for a city; its operations and maintenance expenses would be paid for out of passenger fares; but, in all likelihood, these expenses would be subsidized by city taxes. Therefore, the use of a PRT might not constitute a "fair" test of its "market" acceptance.

Private industry is often reluctant to divulge its research costs for a number of reasons, not the least of which is to hide its bad investments. Neither Westinghouse nor General Electric choose to comment publicly on reports that their losses on the turnkey reactor projects approaches a billion dollars.[6] Furthermore, for atomic reactors, the relative costs of nuclear-generated electricity is partly a function of nonnuclear alternatives (such as coal, gas, and oil). Nuclear power might be economically inefficient at a given price compared with electricity generated by fossil fuel, but with no improvement in nuclear technology, it suddenly becomes economically attractive because of a rise in other fuel costs.

For nuclear reactors, this objective might be measured by the number of commercial-sized reactors built, but this still overlooks the problem of government subsidies. Perhaps a better measure would be the cost of nuclear electricity relative to other means of generating electricity upon reaching commercial-sized reactors (at some defined point in either time or development stage) or the percentage of nuclear electricity relative to other forms within a given nation, either in operation or projected. A third measure could be the ratio of planned nuclear power plants to conventional power plants. This would also provide an indication of what utility decisionmakers see as future trends in

---

[6] The figure is from Perry et al., *The Light Water Reactor*, p. 38.

relative costs and technology efficiency. These all suggest applying a crude market test to the nuclear production of electricity, based upon the assumption that a nation's electric utilities would be reluctant to rely on nuclear power for generating electricity if it were consistently perceived as more expensive or less reliable than other sources.

If one purpose of developing an efficient technology were to capture an export market, the most straightforward measure for such an objective might be the number of units sold abroad, such as the number of Concordes or nuclear reactors. This would reflect the success with which a nation achieved its targeted export market. There are three fundamental problems with this measure, particularly in the case of nuclear reactors. First, for the Soviet Union, exports to a Communist bloc nation from the Soviet Union can hardly be equated with the export of a similar commodity from West Germany to Brazil. Second, in many cases there are sizable subsidies (or other favorable conditions) that a national government is willing to provide to persuade a potential purchaser to buy that nation's technology, thus effectively undermining the open market assumption that the number of units sold implies. For example, West Germany offered to sell a plutonium reprocessing facility to Brazil (and France to Pakistan) as an added inducement to purchase a West German reactor system, an incentive no other supplier was willing to add. Foreign licensing arrangements also complicate this issue. Third, and most damaging, the amount of goods sold abroad does not accurately reflect the intensity to which the export objective was a driving factor, even though it does suggest the attainment of that objective. For instance, the British and the French clearly predicated their development of the Concorde on sales to foreign airlines (especially Pan American and TWA); it was obvious to all that the British and French governments had no hope of regaining their investment if their SST sales were restricted to their domestic markets. That no Concordes were exported does not reflect the magnitude of the Anglo-French efforts, nor does it give an accurate reading as to the importance of that objective in motivating the development.[7]

---

[7] These issues are central to Costello and Hughes, *The Concorde Conspiracy*.

Probably the only accurate measure of this objective would be to estimate the number of government trade commissions and delegations and (where applicable) the private industry counterparts that labored to sell a given technology in an external market. However, these figures would be impossible to obtain or to attribute much credibility to, even if the practice of large bribes and other perquisites did not exist.

The most available means for currently determining some approximation of this objective is to draw careful inferences from public statements on this subject. A second possibility would be to gauge the domestic demand for a given technology—that is, to estimate the capacity of the domestic market to support the fully developed technology without the benefit of external sales. For example, British atomic energy officials were quite open about their desire to develop nuclear reactors that could be sold abroad; furthermore, given the number of industrial consortia formed to produce nuclear reactors and the projected increase in English electricity demands, it was clear that reactors would have to be sold abroad if the nuclear technology industry was ever to be free of heavy subsidization by the government. Similar arguments can be cited for the British-French development of the Concorde.

Political Equity

This objective refers to the activities of various interest groups and the government's response to their often contradictory demands. These interest groups could represent a wide range of clients, including scientific, industrial, utility, labor, and perhaps even consumer lobbies. A set of such groups is present in every government-sponsored development program; if they were not, there would be little incentive for government funding given this depiction of the government as a reactive distribution system.

Unfortunately, at this time, we must rely on accumulating individual case studies detailing the interactions of these groups and the success or failures of each. Their strategies (or at least those that can be observed) suggest what demands are placed upon the government. The governmental decisionmaking process implies how the government arrives at some determination of political equity. However rough the measure might be, the allocation of funds gives some indication of the effectiveness of the demands. Some government allocation decisions serve multiple interest groups. For example, an UMTA grant for a rapid transit system could be seen as a positive, simultaneous response to municipal, labor, and aerospace lobbies. This multiple benefit aspect renders rather problematic judgments about "who gets what." For these reasons, the problems of measuring political equity calculations must remain idiosyncratic to the different cases until a sufficiently large number of comparable technology development cases can be aggregated and a typology developed.[8] Until then, we must use such rough indicators as the initiation of government support of demonstration projects as evidence of the political equity processes. Even this measure has its difficulties, in that it does not give full recognition to the intragovernmental mechanisms. That is, what direct and indirect effects do various government bureaucracies have on the political equity processes?

Attributes

Active Scientific Community

This attribute is meant to capture the role of a nation's scientific community and its supporters in promoting a government's involvement in and financial support of large-scale technology developments. It might best be approximated through an examination of various cases of technology development and determining the scientific and political roles of the scientific community.

Integrated Technology Delivery System

The means for measuring and assessing this attribute describing the relationship among the major institutional actors in the

[8] See Lowi, "Four Systems of Policy, Politics, and Choice," for an example of such a typology.

development and dissemination of a technology must likewise arise out of a number of individual case studies. The need for a broad range of studies is especially critical if the TDS concept is to prove valuable for cross-national comparison purposes. Specifically, one needs to determine if during a given technology's development the posited three major organizational sectors or actors existed and, if so, were they at odds with each other or were their relationships productive and harmonious? This judgment should not be taken at any specific point but over the entire development experience.

Concomitant Military Program

The measurement of this attribute is fairly clear-cut. Did the civilian technology programs gain any "spillover" effects from a concomitant national military program? Spinoffs from space technology research also could be considered here. In some cases, the necessary nexus might be difficult to identify; for example, does aerospace technology directly influence the design of PRT vehicles or biological warfare research benefit medical research? Still, in most cases, the existence of the two parallel programs and the critical linkages that might exist can be established.

Early Starter

The measure of this attribute focuses on the development and dissemination periods rather than the early periods characterized by basic research. Were the early and development periods based largely upon indigenous research that roughly defined the technological frontiers of that technology (was that nation an "early starter"?) or were they based on proven, foreign designs? In most cases, this assessment is easy to make.

R&D Heritage

This variable is meant to capture a nation's "track record" of past performances in large R&D or scientific projects. The critical assessment is the existence of a previously productive R&D establishment in which the specified technology development could be grounded. This attribute might be measured by the size of the nation's total R&D expenditures over a given period of time or the number of important technological innovations produced or patents issued.[9] These would offer a surrogate measure of the nation's tradition in encouraging and utilizing technical R&D.

Multiple Technology Options

This variable can be viewed as considering the number of technological alternatives that a nation's R&D establishments brought as far forward in the RD&D process as the development phase and, even better, to the demonstration stage. The measure for this attribute is quite clear. How many technology options of a specified technology were brought as far as the defined development stage? Where this might be more obscured is in deciding what variations on given technology can be classified as distinct alternatives; for example, do the BWR and PWR represent one (both are light water reactors) or two alternatives?

Resource Investment

This attribute refers to the financial and personnel resources, both government and nongovernment, expended on a given technology development, either as a cumulative expenditure (perhaps per capita to normalize the expenditures) or within developmental stages. To reflect the role of government versus nongovernment expenditures in the different nations, this attribute could be expressed as a ratio of government RD&D spending to total (public and private) RD&D spending for a given technology or government spending to nongovernment spending. This calculation should also include government subsidies for the purchase and perhaps operation of a technology.

---

[9] For one attempt at systematically ordering such data, see OECD, *Gaps in Technology: Comparisons Between Member Countries in Education, Research & Development, Technological Innovation.*

This variable presents pressing data definition and acquisition problems. Among the definitional problems, central is the distinction between government and nongovernment spending, especially outside the United States. British Aircraft Corporation (which built the British version of the Concorde) and BOAC (which bought the aircraft) are, of course, both nationalized industries. This does not even begin to expose the definitional problems when there is no discernible private industry (the Soviet Union being the most obvious case) or when private industry is deliberately structured and managed by the government (as was the case in the British and French developments of the gas-graphite reactors). The distinction between private and public is simply not always available in comparative policy analysis.

The second problem is just as serious. The difficulties of obtaining consistent, comparable time series cost data for technology developments across a number of nations presents enormous obstacles. In a few cases, such figures are available; the costs of the American SST and the Morgentown PRT demonstration projects are matters of public record, and estimates of the Concorde's costs are cited. But in the greater number of cases, cost figures are not available. Reliable cost figures cannot even be obtained for the American nuclear reactor program, let alone the foreign programs, and even if they were available, the problems of comparing purchasing power across societies still loom. For this reason, a number of surrogates must be investigated. For example, in many cases the government atomic energy budget or staff size can be obtained and nongovernment investments can be inferred from a number of published sources. However, most government nuclear agencies engage in both civilian and military research activities that are difficult to separate; many vendors treat their investments as proprietary information; and the problems of separating government from nongovernment spending still arise. In short, this particular variable presents serious data definitional and availability problems. However, to exclude the resources invested would deny a major characterization of technology development to the general model being proposed. Much effort needs to be devoted to developing and refining this variable.

Appendix B: The Mechanics of Nuclear Power Reactors

A nuclear power reactor operates on the simple principle that the splitting of an atomic nucleus of certain elements releases both neutrons and a specified amount of energy. These released neutrons are either absorbed or hit other atoms, producing a "chain reaction," thus generating more energy and more neutrons and continuing the chain reaction. A nuclear power reactor is designed to control the chain reaction and the subsequent release of energy by regulating the rate at which neutrons are absorbed. Most of the energy produced by this phenomenon occurs as heat. A coolant circulates within the pressure vessel or core of the reactor to remove the heat produced; this coolant generates steam, which drives the turbines and generates electricity (see Figure B-1).

We can conceive of four areas of a reactor system that we can use to distinguish the different principal reactor systems referred to in this study: fuel, moderator, primary coolant, and secondary coolant. The different materials used in these areas are arrayed in Table B-1.

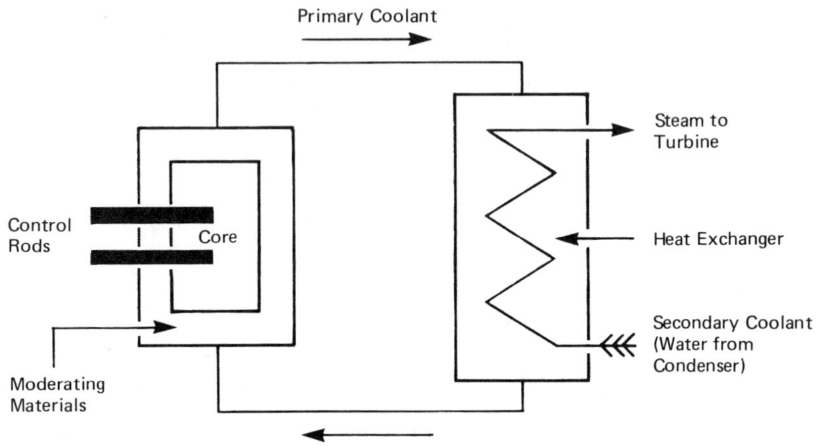

Figure B-1. Schematic Representation of Nuclear Reactor System

(Source: Adapted from Glasstone and Sesonske, *Nuclear Reactor Engineering*, p. 17).

Most of the reactors discussed in the text are fueled by uranium; plutonium is usually conceived of as a future reactor material. The American, Soviet, and German reactor systems were fueled with the uranium-238 isotope that had been enriched with approximately 3 percent of the fissile uranium-235 isotope. The Canadian, British, and French reactor systems were designed around the use of unenriched (often referred to as natural) uranium-238 for fuel. Uranium dioxide pellets are the usual form of fuel.

A number of moderators have been used. The U.S. light water reactors are moderated by circulating water and graphite moderating rods that can be inserted or withdrawn from the core. The Canadian reactors are generally moderated by deuterium oxide (D_2O or "heavy water"), and the initial French and British systems were moderated by graphite rods. The Soviet reactors used both systems.

The primary coolant's tasks are to transfer heat away from the chain reaction occurring in the core and to produce steam for turning the turbines to generate electricity. The U.S. and Cana-

Table B-1. Reactor Materials

Fuel	*Moderator*	*Primary Coolant*	*Secondary Coolant*
Natural uranium	Water	Water	Water
Enriched uranium	Heavy water	Heavy water	
Plutonium	Graphite	Liquid sodium	
	Organic liquids	Organic liquids	
		Gas	

dian systems use light and heavy water, respectively, as the primary coolant. The French and British systems used gas (usually pressurized carbon dioxide). Liquid sodium was used in some early American reactor developments and is the primary cooling medium proposed for the breeder reactor; a number of reactors have experimented with various organic solutions (e.g., diphenyl) as a primary coolant to avoid the corrosive problems of water.

Finally, all the reactors studied here used water as a secondary coolant to reduce the temperature of the steam that drives the turbines so that it can be recirculated back into and cool the reactor core.

Major Reactor Types

This book has focused on four major reactor types: the pressurized water reactor plant; the boiling water reactor plant; the gas cooled, graphite moderated reactor system; and the heavy water moderated and cooled reactor system. As noted, these reactor types are distinguished by their fuels, moderators, and coolants. They are also differentiated by their operations. The following descriptions highlight the basic operational distinctions.[1]

---

[1] For more technical detail, see Glasstone and Sesonske, *Nuclear Reactor Engineering*, ch. 13; or John F. Hogerton, *The Atomic Energy Deskbook* (New York: Reinhold, 1963).

Pressurized Water Reactor (PWR)

The primary coolant in a pressurized water reactor is maintained under pressure, thereby permitting it to leave the core at a higher temperature than obtained with the boiling water reactor (see Figure B-2). The coolant is then sent to a heat exchanger where it heats water to steam temperature, which drives the turbines (high and low pressure turbines in tandem to capture as much energy as possible). The pressurized water, having transferred most of its heat, is recirculated through the reactor while the steam is cooled by the secondary coolant and returned to the heat exchange system.

Boiling Water Reactor (BWR)

The boiling water reactor permits the water in the primary cooling system to boil. This produces steam, which is dried and channeled to the turbines that generate electricity. The steam is then cooled by the secondary coolant and returned to the reactor pressure vessel as water (see Figure B-3). The BWR eliminates the heat exchange mechanism of the PWR and has more modest coolant circulation requirements.

Gas-Graphite Reactor (GGR)

The gas cooled reactor was the primary reactor system developed in France and Britain but was also seriously considered in the United States in 1943. Helium was the preferred coolant proposed for the Hanford reactor, but the expensive helium and expected leakage problems prevented using the gas as a coolant and resulted in the substitution of Columbia River water. In the United Kingdom, air was used as the coolant for the Windscale reactor because the location lacked an adequate water source and because air was thought to be superior from a safety standpoint. Nevertheless, gas leakage problems were severe during the early days; in Calder Hall, the initial leakage rate was two tons of CO_2 per day for each reactor.

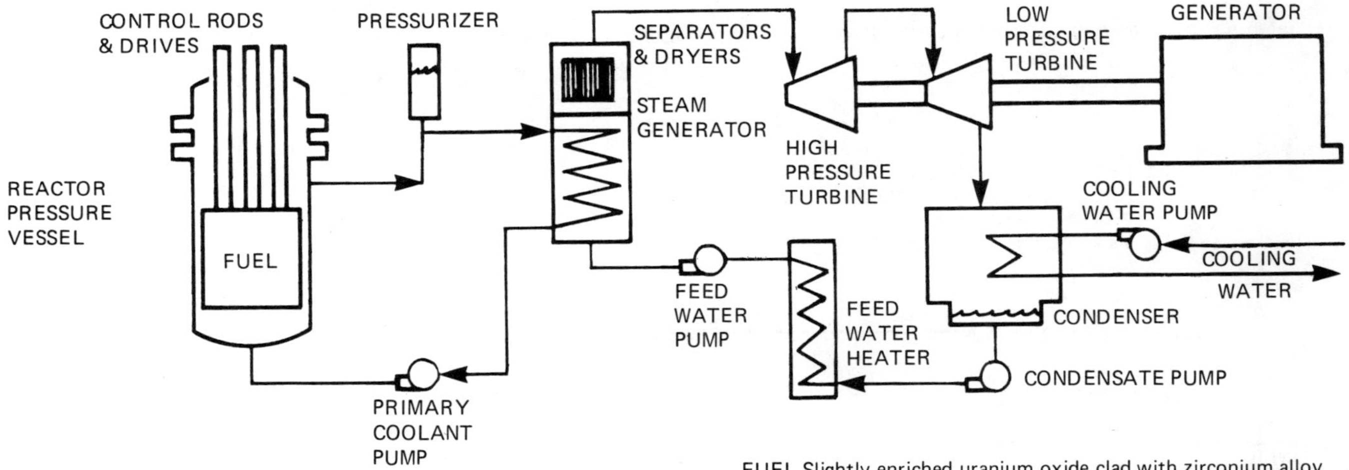

FUEL Slightly enriched uranium oxide clad with zirconium alloy
MODERATOR Water
COOLANT Water
PRESSURE OF PRIMARY SYSTEM 2.250 pounds per square inch
REACTOR OUTLET TEMPERATURE 605° F

Figure B-2. Pressurized Water Reactor Power Plant

(Source: U.S. Atomic Energy Commission, *Power Plant Capital Costs—Current Trends and Sensitivity to Economic Parameters*, WASH-1345, Washington, D.C., October 1974).

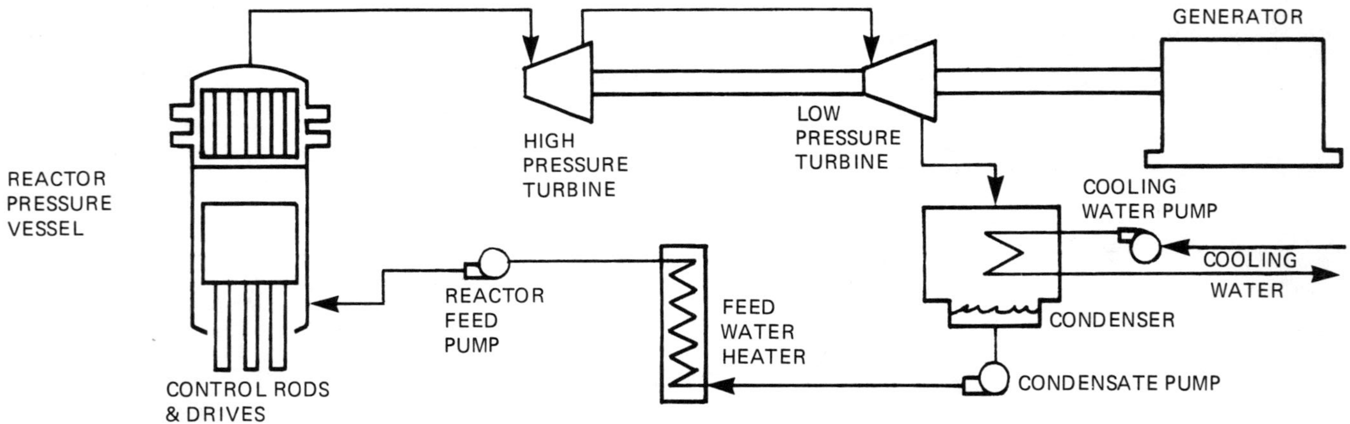

FUEL Slightly enriched uranium oxide clad with zirconium alloy
MODERATOR Boiling water
COOLANT Boiling water
PRESSURE OF PRIMARY SYSTEM 1,000 psi
OUTLET TEMPERATURE 550° F

Figure B-3. Boiling Water Reactor Power Plant

(Source: U.S. Atomic Energy Commission, *Power Plant Capital Costs—Current Trends and Sensitivity to Economic Parameters*, WASH-1345, Washington, D.C., October 1974).

The operation of a gas cooled, graphite moderated reactor is schematically similar to the pressurized water reactor (Figure B-2) with two important changes. First, the British and French reactor systems used graphite to moderate their reactors; and second, carbon dioxide under pressure (150 to 250 psi) was used to transfer the reactor heat to the heat exchange unit. Special finned designs of various types for fuel elements were developed to facilitate heat removal from the core. In the heat exchange unit, the heated gas would raise the temperature of water to steam, which would then be used to turn the electric turbines; the chemical properties of the reactor's components limited the possible steam temperatures to 650°F, which is fairly inefficient for generating electricity. The selection of gas as a coolant was largely determined by the choice of unenriched uranium as the reactor fuel.

The advanced gas reactor (AGR) raises the temperature of the CO_2 to a higher and therefore more efficient temperature for producing steam. Slightly (1.2 percent) enriched uranium is used; the CO_2 is pressurized to 270 psi, which permits the steam conditions to be raised to 850°F.

Heavy Water Reactor (HWR)

Like the GGR, the heavy water reactor was designed to operate on natural uranium and is similar schematically to the PWR (see Figure B-2). The heavy water primary coolant is maintained under pressure (500 to 1,000 psi) and used to produce saturated steam. There are some important differences, however. First, and the most obvious, heavy water is used as both the moderator and the primary coolant. Second, in the CANDU system, the pressure tubes are horizontal and the fuel elements are inserted into the core from both directions. Because natural uranium is used for fuel, the neutron economy characteristics of heavy water are necessarily reinforced by cladding the fuel elements in zirconium alloys.

Appendix C: Chronology of Major National "Slow" Reactor Programs

		Development Stage	Demonstration Stage	Dissemination Stage	
United States					
PWR	1947	Shippingport (1951)	Yankee-Rowe (1961)	Yankee-Haddam (1968)	
BWR	1947	Vallecitos (1967)	Dresden (1960)	Oyster Creek (1969)	
West Germany					
PWR	1955		Obrigheim (1969)	Stade (1972)	
BWR	1955	Kahl (1962)	Grundremmingen (1966)	Würgassen (1972)	
Soviet Union					
LWGR	1949	Obinsk (1954)	Beloyarsk 1-2 (1954-1967)	Beloyarsk 3-4 (1971-1972)	
PWR	1949	"Lenin" (1956)	Novovoronezh 1-2 (1954-1969)	Novovoronezh 3-4 (1971-1972)	
BWR	1949	Melekess (1962)			
Canada					
HWR	1945	NPD (1962)	Douglas Pt. (1967)	Pickering 1-4 (1971-1973)	
France					
GGR	1949	Marcoule G2, G3 (1959-1960)	Chinan 2-3 (1965-1967)	St. Laurent 1-2 (1969-1971)	1968
PWR			Chooz (1967)	Fessenheim 1-2 (1976)	
England					
GGR	1946	Calder Hall (1956)	Magnox Reactor (1962-71)	1965	
AGR		Windscale (1963)	Dungeness B (1976) 1974		
SGHWR		Winfreth (1967)	1978		

PWR = Pressurized Water Reactor
BWR = Boiling Water Reactor
LWGR = Light Water Graphite Reactor
HWR = Heavy Water Reactor
SGHWR = Steam Generating HWR
GGR = Gas Graphite Reactor
AGR = Advanced Gas Reactor

References

"The Advanced Gas Cooled Reactor." *Nuclear Engineering* 6, 59 (April 1961):151-157.

Alexander, A.J. *R&D in Soviet Aviation*. Santa Monica, Calif.: The Rand Corporation, R-589-PR, November 1970.

________. "Weapon Acquisition in the Soviet Union, United States, and France." In *Comparative Defense Analyses*, edited by F.B. Horton II, A.C. Rogerson, and E.L. Warner, III, pp. 426-433. Baltimore: The Johns Hopkins Press, 1973.

Allen, W. *Nuclear Reactors for Generating Electricity: U.S. Development from 1946 to 1963*. Santa Monica, Calif.: The Rand Corporation, R-2116-NSF, July 1977.

Allison, G.T. *Essence of Decision: Explaining the Cuban Missile Crisis*. Boston: Little, Brown, 1971.

Almond, G.A. "Public Opinion and the Development of Space Technology. In *Outer Space in World Politics*, edited by J.M. Goldsden. New York: Praeger, 1963.

Amann, R.; Berry, M.J.; and Davis, R.W. "Science and Industry in the USSR." In *Science Policy in the USSR*, edited by E. Zaleski et al., pp. 377-582, Paris: Organisation for Economic Co-operation and Development, 1969.

Armand, L.; Etzel, F.; and Giordani, F. *A Target for Euratom*. European Economic Community, May 1957.

Ashford, D.E. "Political Science and Policy Studies: Towards a Structural Solution." *Policy Studies Journal* 5, Special Issue (June 1977):570-583.

Aspen, L. "How to Look at the Soviet-American Balance." *Foreign Policy* 22 (Spring 1976):96-106.

Atomic Energy of Canada, Ltd. *Annual Report 1974-75.* Ottawa: 1975.

"Atomic Energy Control Board." *Nuclear Engineering* 13, 145 (June 1968):504.

AURUM. "Britain Chooses a Reactor." *Bulletin of the Atomic Scientists* 28, 2 (February 1972):45.

Baer, W.S.; Johnson, L.L.; and Merrow, E.W. *Analysis of Federally Funded Demonstration Projects: Final Report.* Santa Monica, Calif.: The Rand Corporation, R-1926-DOC, April 1976.

________. "Government-Sponsored Demonstrations of New Technology." *Science* 196, 4293 (May 27, 1977):950-957.

________. *Analysis of Federally Funded Demonstration Projects: Supporting Case Studies.* Santa Monica, Calif.: The Rand Corporation, R-1927-DOC, April 1976.

Baker, S.J. "Nuclear Proliferation: Monopoly or Cartel?" *Foreign Policy* 23 (Summer 1976):202-220.

Bauer, R.; Poole, I. deS.; and Dexter, L. *American Business and Public Policy.* New York: Atherton Press, 1962.

Beaton, L., and Maddox, J. *The Spread of Nuclear Weapons.* London: Chatto and Windus for the Institute for Strategic Studies, 1962.

Belnat, D.F. "Brazil's Nuclear Program Aimed at Filling Energy Gap." *Los Angeles Times* (May 30, 1977), section I, pp. 1, 6.

Berliner, J.S. *The Innovative Decision in Soviet Industry.* Cambridge, Mass.: The MIT Press, 1976.

"Better Basis for German Third Program." *Nuclear Engineering* 13, 142 (March 1968):159.

"A Better Calder for the World Market." *Nucleonics* 14, 12 (December 1956):56-59.

Brewer, G.D. *Politicians, Bureaucrats, and the Consultant.* New York: Basic Books, 1973.

"British Debates for Atomic Power." *Bulletin of the Atomic Scientists* 9, 4 (May 1953):141-144.

"British Government Speeds Nuclear Program." *Nucleonics* 14, 12 (December 1956):53-55.

Brooks, J. *The Fate of the Edsel and Other Business Adventures.* New York: Harper and Row, 1959.

Bupp, I.C., and Derian, J.C. *Light Water: How the Nuclear Dream Dissolved.* New York: Basic Books, 1978.

________. "The Nuclear Power Industry." In *The Management of Global Issues,* Appendix B of *The Commission on the Organization of the Government for the Conduct of Foreign Policy*, vol. 1. Washington, D.C.: U.S. Government Printing Office, June 1975, pp. 85-100.

Burgess, H.L.J. "Nuclear Engineering in Germany." *Nuclear Engineering* 5, 46 (March 1960):113-115.

Burn, D. *The Political Economy of Nuclear Energy.* London: Institute for Economic Affairs, 1967.

Carle, R., and Gibrat, R. "EL4 and the French Heavy Water/Gas Line." *Nuclear Engineering* 10, 108 (May 1965):171-174.

Carruthers, J. "Canada Finally Clinches Sales of Two Reactors to Romania." *Energy Daily* 6, 202 (October 19, 1978):4.

Carter, J. *Why Not the Best?* New York: Bantam, 1976.

Cartwright, H. "SGHWR: A Darkhorse Competitor." *Nucleonics* 24, 9 (September 1966):60-63.

Caty, G., et al. *The Research System: Comparative Survey of the Organisation and Financing of Fundamental Research*, vols. 1-3. Paris: Organisation for Economic Co-operation and Development, 1974.

Central Electricity Generating Board. "An Appraisal of the Technical Aspects of Dungeness 'B'." *Nuclear Engineering* 10, 112 (September 1965):144-146.

"Changing Trends in France." *Nuclear Engineering* 12, 139 (December 1967):882.

Cockburn, A "The Nuclear Disaster They Didn't Want to Tell You About." *Esquire* 89, 7 (April 25, 1978):39-43.

Commissariat a l'Energie Atomique. *Rapport Annuel 1970*. Paris: CEA, 1971.

"Committee Probes CEA's Future as EdF Gets Enrichment Go-Ahead." *Nuclear Engineering* 15, 165 (February 1970):76-77.

Costello, J., and Hughes, T. *The Concorde Conspiracy*. New York: Charles Scribner, 1976.

Coulam, R.F. *Illusions of Choice: The F-111 and the Problems of Weapons Acquisition Reform*. Princeton: Princeton University Press, 1977.

Creutz, E. "Nuclear Power: Rise of an Industry." *Bulletin of the Atomic Scientists* 25, 6 (June 1970):75-82.

Crozier, M. *The Bureaucratic Phenomenon*. Chicago: University of Chicago Press, 1964.

Cyert, R.M., and March, J.G. *A Behavioral Theory of the Firm*. Englewood Cliffs, N.J.: Prentice-Hall, 1963.

Cyr, A., and deLeon, P. "Comparative Policy Analysis." *Policy Sciences* 6, 4 (December 1975):375-384.

Dawson, F.G. *Nuclear Power Development and Management of a Technology*. Seattle: University of Washington Press, 1976.

deLeon, P. "Public Policy and Political Development: The Case of the SSRC." *Policy Studies Journal* 5, Special Issue (June 1977):596-615.

Deutsch, K.W. "Cracks in the Monolith: Possibilities and Patterns of Disintegration in Totalitarian Systems." In *Totalitarianism*, edited by C.J. Friedrich. Cambridge, Mass.: Harvard University Press, 1954.

DeYoung, K. "Latin Americans Hurry to Catch Up in Nuclear Power." *Washington Post* (July 8, 1977), section D, pp. 1, 5.

Donnelly, W.H. *Commercial Nuclear Power in Europe: The Interaction of American Diplomacy with a New Technology*. Washington, D.C.: U.S. Government Printing Office, December 1972, prepared for the Subcommittee on National Security Policy and Scientific Developments of the Committee on Foreign Relations, U.S. House of Representatives.

"Dragon." *Nuclear Engineering* 5, 50 (July 1960):299-316.

Duffy, G. "Soviet Export Policy." *International Security* 3, 3 (Winter 1978).

Eads, G. "U.S. Government Support for Civilian Technology: Economic Theory versus Political Practice." *Research Policy* 3, 1 (April 1974):3-16.

_______, and Nelson, R.R. "Government Support of Advanced Civilian Technology: Power Reactors and the Supersonic Transport." *Public Policy* 19, 3 (Summer 1971):405-427.

Editorial. "The Anti-Nuclear Programme." *Nuclear Engineering* 9, 96 (May 1964):151-152.

Editorial. "A Commercial Policy for the AEA." *Nuclear Engineering* 7, 77 (October 1962):381-382.

Editorial. "Commercial Power Stations." *Nuclear Engineering* 2, 10 (January 1957):39.

Editorial. "Couture Cures Confuse." *Nuclear Engineering International* 14, 153 (February 1969):72.

Editorial. "Export Effort Inadequate." *Nuclear Engineering* 3, 25 (April 1958):75-77.

Editorial. "Export Policy." *Nuclear Engineering* 1, 1 (April 1955):2.

Editorial. "Germany." *Nuclear Engineering* 10, 112 (September 1965): 317.

Editorial. "Nu-Can CANDU." *Nuclear Engineering* 13, 145 (June 1968): 497.

Editorial. "Specious Selection." *Nuclear Engineering* 10, 109 (June 1965):108-109.

Eggelston, W. *Canada's Nuclear Story*. Toronto: Clarke, Irwin and Co., 1965.

El Baroudi, M.Y., and Keeping, P.A.T. "A Canadian Initiative in the Uranium Enrichment Field." *Nuclear Engineering International* 19, 217 (June 1974):508-509.

Emelyanov, V.S. "Nuclear Energy in the Soviet Union." *Bulletin of the Atomic Scientists* 27, 9 (November 1971):38-41.

Etzioni, A., and Remp, R. "Technological 'Shortcuts' to Social Changes." *Science* 175, 4017 (January 6, 1972):31-38.

Ezra, A.A. "Technology Utilization: Incentives and Solar Energy." *Science* 187, 4187 (February 28, 1975):707-713.

Finke, Dr. W. "Survey of F.R.G.: Research Activities Directed Towards the Nuclear Generating Programme." *Nuclear Engineering* 13, 148 (September 1968):743-744.

"France: Industry—Its Organization and Associations." *Nuclear Engineering* 8, 80 (September 1963):305-306.

["Frascati Manual."] *The Measurement of Scientific and Technical Activities*. Paris: Organisation for Economic Co-operation and Development, 1976.

Friedman, M. *Essays in Positive Economics*. Chicago: University of Chicago Press, 1963.

Galloway, J. *The Politics and Technology of Satellite Communications*. Lexington, Mass.: Lexington Books, 1972.

Gandara, A. *Utility Decisionmaking and the Nuclear Option*. Santa Monica, Calif.: The Rand Corporation, R-2148-NSF, July 1977.

"GEC Wins British Nuclear Industry." *Nuclear Engineering International* 18, 203 (April 1973):276.

George, A.L., and George, J.L. *Woodrow Wilson and Colonel House*. New York: John Day, 1956.

"German Atomic Energy Act." *Nuclear Engineering* 5, 44 (January 1960):4.

"Germany." *Nuclear Engineering* 5, 55 (December 1960):543.

"Germany." *Nuclear Engineering* 6, 66 (November 1961):474-475.

"Germany: A Review." *Nuclear Engineering* 7, 78 (November 1962):441-445.

Gerwin, R. *Atoms in Germany*. Düsseldorf: Econ-Verlag for the Bundisiminister für Wissenschaftliche Forschung, 1964.

Gibbs, K.P., and Fair, D.R.R. "The Magnox Stations: A Success Story." *Nucleonics* 24, 9 (September 1966):48-53.

Gillette, R. "India: Into the Nuclear Club on Canada's Shoulders." *Science* 184, 4141 (June 7, 1974):1053-1055.

_______. "Nuclear Power in the USSR: American Visitors Find Surprises." *Science* 173, 5001 (September 10, 1971):1003-1006.

_______. "Russ Atom Accident Laid to Reactor." *Los Angeles Times* (November 10, 1976), pp. 1, 10.

Gilpin, R. *France in the Age of the Scientific State*. Princeton: Princeton University Press, 1968.

_______. "The Politics of Transnational Economic Relations." In *Transnational Relations and World Politics*, edited by R.O. Keohane and J.S. Nye. Cambridge, Mass.: Harvard University Press, 1972.

_______. "Technological Strategies and National Purpose." *Science* 169, 3944 (July 31, 1970):441-448.

Glasstone, S., and Sesonske, A. *Nuclear Reactor Engineering*. Princeton: Von Nostrand, 1963.

Goldschmidt, B. *The Atomic Adventure: Its Political and Technical Aspects*, translated by P. Beer. New York: Macmillan, 1964.

_______. "The French Atomic Energy Program." *Bulletin of the Atomic Scientists* 18, 7 (September 1962):39-42; and 18, 8 (October 1962):46-48.

Gowing, M. *Britain and Atomic Energy, 1939-1945*. New York: Macmillan, 1964.

_______. *Independence and Deterrence: Britain and Atomic Energy 1945-1952*, vol. 1. St. Martin's Press, 1967.

Gray, J.L. "AECL in Canada's Nuclear Power Programme." *Nuclear Engineering* 13, 145 (June 1968):498-501.

_______. "The CANDU-PHW Programme." *Nuclear Engineering International* 19, 217 (June 1974):476-477.

_______. "Frontiers of Technology: Canada." *Nuclear Engineering* 9, 100 (September 1964):315-316.

Green, H.P., and Rosenthal, A. *Government of the Atom: The Integration of Powers*. New York: Atherton, Press, 1963.

Greenberg, D.S. *The Politics of Pure Science*. New York: New American Library, 1967.

Greenberg, G.D. et al. "Developing Public Policy Theory: Perspectives from Empirical Research." *American Political Science Review* 71, 4 (December 1977):1532-1543.

Greenwood, T. *Making the MIRV: A Study of Defense Decision Making.* Cambridge, Mass.: Ballinger, 1975.

Groves, L.R. *Now It Can Be Told: The Story of the Manhattan Project.* New York: Harper and Row, 1962.

Gueron, J. "Atomic Energy in Continental Western Europe." *Bulletin of the Atomic Scientists* 26, 6 (June 1970):62-68, 116.

H.A. "The British Program of Nuclear Power." *Bulletin of the Atomic Scientists* 11, 6 (June 1955):224-225.

Harman, A.J. *The International Computer Industry: Innovation and Comparative Advantage.* Cambridge, Mass: Harvard University Press, 1971.

Hawkes, N. "Britain Opting for U.S.-Style Reactors—Maybe." *Science* 199, 4430 (February 17, 1978):755-756.

Haywood, L.R. "Heavy-Water Production." *Nuclear Engineering International* 19, 217 (June 1974):506-507.

Heidenheimer, A.J.; Heclo, H.; and Adams, C.T. *Comparative Public Policy: The Politics of Social Choice in Europe and America.* New York: St. Martin's Press, 1975.

Hewlett, R.G., and Anderson, O.E. *A History of the United States Atomic Energy Commission: The New World, 1939-1946*, vol. 1. University Park: Pennsylvania State University Press, 1962.

_______, and Duncan, F. *A History of the United States Atomic Energy Commission: Atomic Shield, 1947-1952*, vol. 2. University Park: Pennsylvania State University Press, 1969.

_______. *Nuclear Navy 1946-1962.* Chicago: University of Chicago Press, 1974.

Hilberry, N. "Nuclear Power in the U.S." *Nuclear Industry* 11, 8 (August 1964):5-12.

Hill, G.C. "Troubled Transit: Some Experts Think 'Personal Transit' Is Wave of the Future." *Wall Street Journal* (August 30, 1974), pp. 1, 19.

Hinton, C. "Atomic Power in Britain." *Scientific American* 198, 3 (March 1958):29-35.

Hirsch, R. "French Nuclear Policy." *Nuclear Engineering* 11, 127 (December 1966):938-939.

Hirschman, A.O. *Exit, Voice, and Loyalty.* Cambridge, Mass.: Harvard University Press, 1970.

_______. "The Search for Paradigms as a Hindrance to Understanding." *World Politics* 22, 3 (April 1970):329-343.

Hodgetts, J.E. *Administering the Atom for Peace.* New York: Atherton Press, 1964.

Hogerton, J.T. "The Arrival of Nuclear Power." *Scientific American* 218, 2 (February 1968):21-31.

_______. *The Atomic Energy Deskbook.* New York: Reinhold, 1963.

Holt, J.O. "Canadian HWR: Economic Assessment." *Nuclear Engineering* 11, 121 (June 1966):461-463.

Holt, R.T., and Turner, J.E. "Crises and Sequences in Collective Theory Development." *American Political Science Review* 69, 3 (September 1975):979-994.

Ibrahim, Y.M. "U.S. Nuclear Industry Sees Bill as Threat to Exports." *New York Times* (February 15, 1978), pp. 51, 57.
Irving, D. *The Virus House: The German Atomic Bomb Project*. New York: Simon and Schuster, 1968.
Jay, K. *Calder Hall*. London: Methuen, 1961.
Jensen, W.G. *Nuclear Power*. Cambridge, Eng.: G.T. Foulis, 1969.
Johnson, L.L. et al. *Alternative Institutional Arrangements for Developing and Commercializing Breeder Reactor Technology*. Santa Monica, Calif.: The Rand Corporation, R-2069-NSF, November 1976.
Joskow, P.L. "The International Nuclear Industry Today." *Foreign Affairs* 54, 4 (July 1976):788-803.
________. "Research and Development Strategies for Nuclear Power in the United Kingdom, France, and Germany." In *Proceedings of the Workshop on Institutional Alternatives for LMFBR Development and Commercialization*, pp. 227-328. McLean, Va.: The MITRE Corporation, 1977.
Julia, R. "French Industry Faces the Challenge of Nuclear Engineering." *Nuclear Engineering* 11, 127 (December 1966):943-949.
Kaiser, R.G. *Russia: The People and the Power*. New York: Athaeneum, 1976.
Kaplan, A. *The Conduct of Inquiry: Methodology for Behavioral Science*. San Francisco: Chandler, 1964.
Keck, O. "West German Science Policy Since the Early 1960's: Trends and Objectives." *Research Policy* 5, 2 (April 1976):115-157.
Kervern, G.Y. "Industrial Production of Nuclear Grade Materials." *Nuclear Engineering International* 17, 196 (September 1972):707-709.
Kilborn, P.T. "British Commission Urges a Delay in Widening Use of Nuclear Energy." *New York Times* (September 23, 1976), pp. 1, 5.
Kildow, J.T. *INTELSAT: Policy-Maker's Dilemma*. Lexington, Mass.: Lexington Books, 1972.
Kissinger, H.A. *The Troubled Partnership*. New York: McGraw-Hill for the Council on Foreign Relations, 1965.
Knorre, H. "Frontiers of Technology: U.S.S.R." *Nuclear Engineering* 9, 100 (September 1964):317-318.
Koopmanns, T. *Three Essays on the State of Economic Science*. New York: McGraw-Hill, 1957.
Kramish, A. *Atomic Energy in the Soviet Union*. Palo Alto: Stanford University Press, 1959.
________. "Atomic Energy in the USSR." *Bulletin of the Atomic Scientists* 15, 8 (October 1959):322-328.
Kuhn, T.S. *The Structure of the Scientific Revolutions*. Chicago: University of Chicago Press, 1970.
Kuhnel, R. "The 15 MW BWR at Kahl." *Nuclear Engineering* 6, 57 (February 1961):56-65.
Kurth, J.R. "Why We Buy the Weapons We Do." *Foreign Policy* 11 (Summer 1973):33-56.
Lambright, W.H. *Governing Science and Technology*. New York: Oxford University Press, 1976.
________, and Teich, A.H. "Technology Transfers as a Problem in Interorganizational Relationships." *Administration and Society* 8, 1 (May 1976):29-54.

Laurence, G.C. "NRU Operating." *Nuclear Engineering* 3, 22 (January 1958):19-25.
Law, C.A. "Heavy Water Production." *Nuclear Engineering* 13, 145 (June 1968):510-512.
"Lenin—The Russian Ice-Breaker." *Nuclear Engineering* 2, 31 (October 1958):432-433.
Leny, J.C. "ORGEL—A European Concept." *Nuclear Engineering* 6, 67 (December 1961):508-512.
Lewis, W.B. "The Accident to the NRX Reactor on December 12, 1952." Toronto: AECL Paper No. 232, 1953.
_______. "An Atomic Power Proposal." DR-18, Toronto: AECL Paper No. 186, DR-18, August 1951.
_______. "Canada's Steps Toward Nuclear Power." In *Progress in Nuclear Energy-Economics*, London: Pergamon Press, vol. 2, pp. 227-288.
Lindsey, R. "U.S. May Raze Its $57-Million Showcase 'People-Mover'." *New York Times* (April 13, 1974), pp. 1, 26.
Loftness, R.L. *Nuclear Power Plants*. New York: Van Nostrand, 1964.
Logsdon, J.M. *The Decision to Go to the Moon: Project Apollo and the National Interest*. Cambridge, Mass.: The MIT Press, 1970.
Long, T.D., and Wright, C., eds. *Science Policies of Industrial Nations*. New York: Praeger, 1975.
Lowi, T.J. "Decision Making vs. Policy Making: Towards an Antidote for Technocracy." *Public Administration Review* 30, 3 (May/June 1970):314-325.
_______. "Four Systems of Policy, Politics, and Choice." *Public Administration Review* 32, 4 (July/August 1972): 298-310.
Lowranc, W.W. "Nuclear Futures for Sale: To Brazil from West Germany." *International Security* 1, 2 (Fall 1966):147-166.
MacDonald, D.S. "Canada's Role in World Developments of Nuclear Energy." *Nuclear Engineering International* 19, 217 (June 1974):475.
MacKenthun, W. "Survey of Germany: Electric Supply Industry." *Nuclear Engineering* 13, 148 (September 1968):745-748.
MacLachlan, A. "The Russians Are Coming . . . On Strong in Nuclear." *Energy Daily* 6, 200 (October 17, 1978):3-4.
_______. "Vive La France Nucleaire!" *Energy Daily* 6, 199 (October 16, 1978):3-5.
"Magnox Stations Lead in Electricity Production." *Nucleonics* 24, 3 (March 1966):48-56.
Majone, G. "Technology Assessment and Policy Analysis." *Policy Sciences* 8, 2 (June 1977):173-175.
Mansfield, E. *The Economics of Technological Change*. New York: Norton, 1968.
March, J.G. "The Business as a Political Coalition." *Journal of Politics* 24, 4 (November 1962):662-678.
_______, and Simon, H.A. *Organizations*. New York: John Wiley, 1958.
"The Marcoule Reactor G-2." *Nuclear Engineering* 4, 43 (December 1959):435-437.
Marschak, T.; Glennan, T.K., Jr.; and Summers, R. *Strategies for R&D: Studies in the Microeconomics of Development*. New York: Springer-Verlag, 1967.

Marshall, A.W. "Estimating Soviet Defense Spending." *Survival* 18, 2 (March/April 1976):73-79.
Martino, J.P. *Technological Forecasting for Decisionmaking*. New York: American Elsevier, 1972.
Marweh, O. "India's Nuclear and Space Program: Intent and Policy." *International Security* 2, 2 (Fall 1977):96-121.
Mazlich, B. *Kissinger: The European Mind in American Policy*. New York: Basic Books, 1976.
McDermott, J. "Technology—The Opiate of the Intellectuals." In *Technology and Man's Future*, edited by A.H. Teich. New York: St. Martin's Press, 1972.
McElheny, V.K. "Decisions on Nuclear Power." *Science* 149, 3682 (July 23, 1965):408-409.
McIntyre, H.C. "Natural-Uranium Heavy-Water Reactors." *Scientific American* 233, 4 (October 1975):17-27.
McLin, J.B. *Canada's Changing Policy*. Baltimore: The Johns Hopkins Press, 1967.
Medvedev, Z. "The Two Decades of Dissidence." *New Scientist* 72, 1025 (November 4, 1976):264-267.
Memmel, L. "The Nuclear Industry of the F.R.G." *Nuclear Engineering* 13, 148 (September 1968):748-749.
Mesthene, E.G. *Technology Change*. New York: New American Library, 1970.
Mohr, L.B. "The Concept of Organizational Goal." *American Political Science Review* 67, 2 (June 1973):470-481.
Morison, E.E. *Men, Machines, and Modern Times*. Cambridge, Mass.: The MIT Press, 1966.
Morison, G.M. "CANDU Nuclear Power in Ontario." *Nuclear Engineering International* 19, 217 (June 1974):486-488.
"MZFR." *Nuclear Engineering* 11, 116 (February 1966):28-29.
Nau, H.R. *National Politics and International Technology: Nuclear Reactor Development in Western Europe*. Baltimore: The Johns Hopkins Press, 1974.
Nelson, R.R. "Intellectualizing About the Moon-Ghetto Metaphor: A Study of the Current Malaise of Rational Analysis of Social Problem." *Policy Sciences* 5, 4 (December 1974):375-414.
________; Peck, J.; and Kalachek, E.D. *Technology, Economic Growth, and Public Policy*. Washington, D.C.: The Brookings Institution, 1967.
Nicholson, W. *Microeconomic Theory: Basic Principles and Extensions*. Hinsdale, Ill.: The Dryden Press, 1972.
Novick, S. *The Careless Atom*. Boston: Houghton Mifflin, 1969.
"Nuclear Power Abroad." *Scientific American* 235, 2 (August 1976):44A.
"The Nuclear Power Programme." *Nuclear Engineering* 5, 50 (July 1960):316.
"Nuclear Power in the U.S.S.R." *Scientific American* 219, 7 (July 1968):50.
Oi, W.Y. "The Federal Subsidy of Conventional Mass Transit." *Policy Analysis* 1, 4 (Fall 1975):613-658.
"Ontario Hydro's Nuclear Activities." *Nuclear Engineering* 13, 145 (June 1968):502-503.

Organisation for Economic Co-operation and Development, *Gaps in Technology: Comparisons Between Member Countries in Education, Research and Development, Technological Innovation*. Paris: OECD, 1970.

PEON Reports. Rapports de la Commission Consultative Pour La Production D'Electricite D'Origine Nucleáire. Paris: Ministère de L'Industrie et de la Researche, 1970.

Peierls, R. "Britain in the Atomic Age." *Bulletin of the Atomic Scientists* 26, 6 (June 1970):40-46.

Perrin, F. "Atomic Energy in France." *Bulletin of the Atomic Scientists* 11, 3 (March 1955):92-93.

Perry, R. *Comparisons of Soviet and U.S. Technology*. Santa Monica, Calif.: The Rand Corporation, R-827-PR, June 1973.

Perry, R.L. et al. *Development and Commercialization of the Light Water Reactor, 1946-1976*. Santa Monica, Calif.: The Rand Corporation, R-2180-NSF, August 1977.

________. *Systems Acquisition Strategies*. Santa Monica, Calif.: The Rand Corporation, R-733-PR/ARPA, June 1973.

Petrosyants, A. "The Atom on the Scales of Progress." *New Times* 13 (March 1976):18-19.

________. *From Scientific Search to Atomic Industry*. Danville, Ill.: Interstate, 1975.

________. "Nuclear Energy—A Productive Force." *New Times* (1971):18-20.

"Pickering—The First Year." *Nuclear Engineering International* 17, 103 (June 1972):475-478.

Polach, J.G. "Nuclear Power in East Europe." *East Europe* (May 1968): 3-12.

________. "Nuclear Power in Europe at the Crossroads." *Bulletin of the Atomic Scientists* 25, 8 (October 1969):15-20.

Pon, G.A. *The Fundamentals of Nuclear Power*. Ottawa: AECL, June 1975.

"Power Reactors in Canada: NPD and CANDU Studied." *Nuclear Engineering* 3, 29 (August 1958):334-338.

Pretsch, J. "Recent Trends in the German Nuclear Program." *Nuclear Engineering* 10, 112 (September 1965):323.

Price, D.K. *The Scientific Estate*. Cambridge, Mass.: Harvard University Press, 1965.

"Progress Report—EL4." *Nuclear Engineering* 11, 121 (June 1966):473.

Pryde, P.R., and Pryde, L.T. "Soviet Nuclear Power." *Environment* 16, 3 (April 1974):26-34.

Ribicoff, A.A. "A Market-Sharing Approach to the Nuclear Sales Problems." *Foreign Affairs* 54, 4 (July 1976):763-787.

Rippon, S. "The AGRs Make a Start." *Nuclear Engineering International* 21, 242/243 (April/May 1976):54-57.

________. "Primary Interest—To See a Healthy Nuclear Industry." *Nuclear Engineering International* 18, 204 (May 1973):383.

Robinson, J.A.L. "The CANDU Reactor System: An Appropriate Technology." *Science* 199, 4329 (February 10, 1978):657-664.

Rolph, E. *Regulation of Nuclear Power: The Case of the Light Water Reactor*. Santa Monica, Calif.: The Rand Corporation, R-2104-NSF, June 1977.

Rosenau, J.N. "Pre-theories and Theories of Foreign Policy." In *Approaches to Comparative and International Politics*, edited by R.B. Farrell. Evanston, Ill.: Northwestern University Press, 1966.

Rotherham, L. "Calder Hall Nuclear Power Station." *Science* 25, 3245 (March 8, 1957):452-454.

Royal Commission on Environment Pollution. Sixth Report, *Nuclear Power and the Environment*. London: HMSO, 1976.

Salamon, L.M., and Siegfried, J.J. "Economic Power and Political Influence: The Impact of Industry Structure on Public Policy." *American Political Science Review* 71, 3 (September 1977): 1026-1043.

Salander, C. "Nuclear Research Centres in the Federal Republic of Germany." *Nuclear Engineering International* 18, 291 (February 1973):97-100.

Samuelson, R.J. "Pulling the Plug on Plutonium." *National Journal* 9, 31 (July 30, 1977):1176-1189.

Schattmann, S.E. "Britain's Nuclear Power Program." *Science* 127, 3313 (June 27, 1958):1489.

Scheinman, L. *Atomic Energy in France Under the Fourth Republic*. Princeton: Princeton University Press, 1965.

———. "Nuclear Safeguards, The Peaceful Atom, and the IAEA." *International Conciliation* 572 (March 1969).

———. "Security and a Transnational System: The Case of Nuclear Energy." In *Transnational Relations and World Politics*, edited by R.O. Keohane and J.S. Nye, pp. 276-299. Cambridge, Mass.: Harvard University Press, 1972.

Scherer, F.M. *Industrial Market Structure and Economic Performance*. Chicago, Rand McNally: 1970.

———. *The Weapons Acquisition Process: Economic Incentives*. Cambridge, Mass.: Harvard University Press, 1964.

Schmidt-Küster, W.J., and Popp, M. "The German Nuclear Programme." *Nuclear Engineering International* 18, 201 (February 1973):94-97.

Schroeder, G. "The Economic Reform as a Spur to Technological Progress in the USSR." *Jahrbuch der Wirtschaft Osteuropas*, vol. 2. Munich: Gunter Alzos Verlag, 1971, pp. 343-374.

Schulman, P.R. "Nonincremental Policy Making: Notes Towards an Alternative Paradigm." *American Political Science Review* 69, 2 (December 1975):1354-1370.

"The Second Nuclear Power Programme." *Nuclear Engineering* 9, 96 (May 1964):158-159.

"Second Year of the U.K. AEA." *Nuclear Engineering* 1, 5 (August 1956):31.

Selznick, P. *TVA and the Grass Roots*. New York: Harper and Row, 1966.

"SGHWR: Advanced, Economic, and Simple." *Nuclear Engineering* 11, 121 (June 1966):468-470.

Shabad, T. "Soviet Union Steps Up Installation of Nuclear Power Plants." *New York Times* (January 17, 1977), pp. D1, D5.
"Soviet Dissident Discloses Holocausts." *Aviation Week & Space Technology* 105, 20 (November 15, 1976):20.
Steinbruner, J.D. *The Cybernetic Theory of Decision*. Princeton: Princeton University Press, 1974.
Stewart, J.C.C. "Frontiers of Technology: United Kingdom." *Nuclear Engineering* 9, 100 (September 1964):311-312.
Stolz, J. "Nuclear Future for Electricité de France." *Nuclear Engineering International* 17, 196 (September 1972):688-692.
"The Structure of the [U.K.] Atomic Energy Authority." *Nuclear Engineering* 1, 1 (April 1955):20-21.
"Structure of the British Nuclear Industry." *Nuclear Engineering* 8, 82 (March 1963):96-99.
Thompson, J.T. *Organizations in Action*. New York: McGraw-Hill, 1967.
Toffler, A. *Future Shock*. New York: Random House, 1970.
Torrisi, F. "Recent Structural Reorganization in the French Nuclear Industry." *Nuclear Engineering International* 17, 196 (September 1972):702-704.
Truman, D.B. *The Governmental Process*. New York: Knopf, 1951.
Trumball, R. "Canadian Minister Dismisses Chief of Atomic Agency." *New York Times* (July 8, 1977), section D, pp. 1, 8.
Tucker, R.C. "The Georges' Wilson Reexamined." *American Political Science Review* 71, 2 (June 1977):608-618.
U.S. Atomic Energy Commission. *The Nuclear Industry 1974*, WASH-1174-73. Washington, D.C.: U.S. Government Printing Office, September 1974.
________. *Power Plant Capital Costs—Current Trends and Sensitivity to Economic Parameters*, WASH-1345. Washington, D.C.: U.S. Government Printing Office, October 1974.
________. *Report to Congress, 1960*. Washington, D.C.: U.S. Government Printing Office, January 1961. Also U.S. AEC *Reports to Congress*, 1962 and 1963.
________. *Soviet Power Reactors*, WASH-1175. Washington, D.C.: U.S. Government Printing Office, August 1970.
U.S. Congress, Joint Committee on Atomic Energy. *Report of the Subcommittee on Research and Development on the Five-Year Reactor Development Program Proposed by the AEC*. Washington, D.C.: U.S. Government Printing Office, March 1964.
Utterback, J.M. "Market Structure and Technological Innovation." *Science* 183, 4125 (February 15, 1974):620-626.
Vendryes, G.A. "Superphenix: A Full-Scale Breeder Reactor." *Scientific American* 236, 3 (March 1977):26-35.
Verba, S. "Some Dilemmas in Comparative Research." *World Politics* 20, 1 (October 1976):111-127.
Vladimirov, L. *The Russian Space Bluff*, translated by D. Floyd. New York: Dial Press, 1973.
Wade, N. "European Doubts About Nuclear Energy." *Science* 194, 4261 (October 8, 1976):164.

Walsh, J. "British Choose Own Reactor for Nuclear Power Program." *Science* 185, 4150 (August 9, 1974):511.

———. "Nuclear Energy: New Study Assails British Program." *Science* 156, 3780 (June 9, 1967):1350-1352.

———. "Nuclear Power: France Forges Ahead on Ambitious Plan Despite Critics." *Science* 193, 4250 (July 23, 1976):305-306, 340.

Walske, C. "Nuclear Electric Power and Proliferation of Nuclear Weapon States." *International Security* 1, 3 (Winter 1977):94-106.

Webber, M.M. "The BART Experience—What Have We Learned?" *The Public Interest* 45 (Fall 1976):79-108.

Weinberg, A. "Power Reactors." *Scientific American* 191, 6 (December 1954):33-39.

Weller, J.A. "What Nuclear Power Means to Canadian Industry." *Nuclear Engineering International* 19, 217 (June 1974):500-501.

Wenk, E., Jr. "Technology Assessment in Public Policy: A New Instrument for Social Management of Technology." *Proceedings of the Institute of Electrical and Electronics Engineers* 63, 3 (March 1975):371-379.

"The West German Nuclear Programme Begins to Gather Momentum." *Nuclear Engineering* 4, 36 (March 1959):96.

"Why British Claim Lead in Nuclear Power." *Nucleonics* 14, 12 (December 1956):S1-S12.

"W.H. Zinn Evaluates Russian Reactor Progress." *Nucleonics* 17, 3 (March 1959):104-107.

Wilcynski, J. "Atomic Energy for Peaceful Purposes in the Warsaw Pact Countries." *Soviet Studies* 26, 4 (October 1974):568-590.

Wildavsky, A. *The Politics of the Budgetary Process*. 2nd ed., Boston: Little, Brown, 1974.

"Will CANDU Do for the UK?" *New Scientist* (December 6, 1973):702-703.

Wilson, A. *The Concorde Fiasco*. Baltimore: Penguin Books, 1973.

Wohlstetter, A. "Spreading the Bomb Without Quite Breaking the Rules." *Foreign Policy* 25 (Winter 1976-1977):88-97, 145-170.

———. et al. *Moving Toward Life in a Nuclear Armed Crowd?* Los Angeles: Pan Heuristics, ACDA/PAB-263, April 1976.

Wolf, C. Jr. "The Present Value of the Past." *Journal of Political Economy* 78, 4 (July/August 1970):783-792.

Wonder, E.F. "Decision-Making and Reorganization of the British Nuclear Power Industry." *Research Policy* 5, 3 (July 1976):240-268.

Woodhead, L.W., and Ingolfsrud, L.J. "Performance of Canadian Commercial Nuclear Units and Heavy Water Plants." *Annals of Nuclear Energy*. London: Pergamon Press, 1975.

York, H. "Multiple Warhead Missiles." *Scientific American* 229, 5 (November 1973):18-27.

Ziegler, M.H. "The French Nuclear Industry." *Nuclear Engineering* 5, 55 (December 1955):547-548.

"3000 MW CANDU at Douglas Point." *Nuclear Engineering International* 14, 152 (January 1969):41.

Index

About the Author

Peter deLeon is a Senior Social Scientist at the Solar Energy Research Institute, Golden, Colorado, and an adjunct professor of Public Administration at the University of Southern California. Dr. deLeon is the author of numerous books and articles on various aspects of the policy process, has presented invited testimony before congressional committees, and serves on the editorial board of *Policy Sciences.* He is currently examining the effect of petroleum imports on the American economy and the role of the federal government in supporting technology research and development activities pertaining to solar energy.

The Rand Graduate Institute

PUBLISHED

Goldhamer, Herbert. *The Adviser.* New York: Elsevier-North Holland, Inc., 1978, a Rand Graduate Institute Book.

deLeon, Peter. *Development and Diffusion of The Nuclear Power Reactor: A Comparative Analysis*, Cambridge, Mass.: Ballinger Publishing Company, 1979, a Rand Graduate Institute Book.

SELECTED RAND BOOKS ON POLICY ANALYSIS

Averch, Harvey A., John E. Koehler, and Frank H. Denton. *The Matrix of Policy in the Philippines.* New Jersey: Princeton University Press, 1971.

Fisher, Gene H. *Cost Considerations in Systems Analysis.* New York: American Elsevier Publishing Company, Inc., 1971.

Hirshleifer, Jack, James C. DeHaven, and Jerome W. Milliman. *Water Supply: Economics, Technology, and Policy*. Chicago, Illinois: University of Chicago Press, 1969.

Quade, E.S. *Analysis for Public Decisions.* New York: American Elsevier Publishing Company, Inc., 1975.